D1243095

MAKING ANIMAL MEANING

THE ANIMAL TURN

Series Editor
Linda Kalof

Series Advisory Board
Marc Bekoff, Juliet Clutton-Brock, Nigel Rothfels

MAKING ANIMAL MEANING

Edited by Linda Kalof and
Georgina M. Montgomery

RECEIVED
NOV 1 3 2012
MINNESOTA STATE UNIVERSITY LIBRARY
MANKATO, MN 56002-8419

Michigan State University Press
East Lansing

QL
85
.M2797
2011

Copyright © 2011 by Michigan State University

⊛ The paper used in this publication meets the minimum requirements of ANSI/NISO Z39.48-1992 (R 1997) (Permanence of Paper).

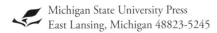 Michigan State University Press
East Lansing, Michigan 48823-5245

Printed and bound in the United States of America.

17 16 15 14 13 12 11   1 2 3 4 5 6 7 8 9 10

LIBRARY OF CONGRESS CATALOGING-IN-PUBLICATION DATA

Making animal meaning / edited by Linda Kalof and Georgina M. Montgomery.
    p. cm. — (The animal turn)
    Includes bibliographical references and index.
    ISBN 978-1-61186-016-0 (cloth : alk. paper) 1. Human-animal relationships. 2. Animals—Psychological aspects. I. Kalof, Linda. II. Montgomery, Georgina M.

    QL85.M2797 2011
    304.2—dc23   2011008826

Cover design by Erin Kirk New

Book design by Scribe Inc. (www.scribenet.com)

Cover Photo: Beluga whale, New York Zoo, 1995. Reproduced with permission of Britta Jaschinski, copyright Britta Jaschinski.

green press INITIATIVE Michigan State University Press is a member of the Green Press Initiative and is committed to developing and encouraging ecologically responsible publishing practices. For more information about the Green Press Initiative and the use of recycled paper in book publishing, please visit www.greenpressinitiative.org.

Visit Michigan State University Press on the World Wide Web at:
www.msupress.msu.edu

Born in the wild waters of Churchill River, Manitoba, Canada, and caught shortly after, this Beluga whale came to New York Aquarium in Coney Island in 1974. For 30 years she circled monotonously, silently in a small chlorine-treated, water-filled concrete tank, till she died in April 2004.

This was her home but she didn't live there! The photo is dedicated to her and all Beluga whales in captivity.

—Photo and Narrative by Britta Jaschinski

Contents

Introduction

Making Animal Meaning explores how humans construct, configure, and constantly negotiate the meaning of other animals in the social world. This meaning-making is not a new human pastime. We have been struggling with the essence of animals for at least 33,000 years—our earliest known surviving artistic endeavours are drawings of lions and rhinoceroses on cave walls in southern France and carvings of birds and horses from mammoth ivory in southwestern Germany. With the onset of writing, the construction of animal meaning took center stage in the first epic poem, *Gilgamesh*, which tells the story of a friendship between a human, King Gilgamesh, and a wild beast-man, Enkidu.

Our attempts to make meaning of animals—to describe their behaviors, depict their unique physical attributes, elucidate their similarities and differences from us, and chronicle our treasured alliances with them—continue unabated to the present day. Contemporary humans still spend much of their time drawing, painting, sketching, and sculpting animal images. We still write their stories and give them central roles in our poetry, fiction, and myths. We are still consumed by the need to mark, unmark, and blur the boundaries between us, creating imaginative human-animal mergers that represent the best, and sometimes the worst, of what it means to be "human" or "animal." And while we have been busily sorting out the meaning of animals, they have been leaving their own traces and signs—thus actively creating their own meaning. Finally, the vast majority of the recent scholarship on animal meaning has been theoretical, offering a stunning array of arguments about the essentials of the "animal," but there is a paucity of empirical research to illustrate the theories of animal essence. This volume begins to fill that gap.

Making Animal Meaning is a collection of ten original essays, three that focus on key theoretical underpinnings of animal meaning and seven that illustrate the theoretical concepts in studies of specific cultural spaces and contexts, animal species, and human-animal relations. Covering some of the most exciting themes in the vibrant field of animal studies, these essays represent the best in interdisciplinarity, including voices from anthropology, science studies, geography, American studies, history, critical studies, environmental studies, and women's studies. The chapters also have significant chronological reach, with essays exploring the Renaissance, the nineteenth and twentieth centuries, and contemporary culture. Together, the chapters reveal animal meanings as they were made in Europe, the United States, Mexico, South Africa, and the Democratic Republic of Congo. A deeply thoughtful and diverse collection, *Making Animal Meaning* ensures the analytical strength of interdisciplinary and international discourse when tackling the age-old quest of animal meaning.

We have divided the book into two sections: Part 1 consists of essays that explore theoretical underpinnings of animal meaning. At the center of these chapters is the animal—the animal as an author who leaves traces of his or her meaning, the animal as a pest who embodies social tensions and conflicts, and the animal as kin with whom we share the act of consumption.

These foundational chapters set the stage for the observational essays in Part 2. Together, these chapters demonstrate new understandings of central questions concerning animal agency, kinship, and consumption. By providing such a forum, *Making Animal Meaning* takes scholars and students into new terrain, reconceptualizing methods for researching animal histories and rethinking the contingency of the human-animal relationship.

In "Animal Writes: Historiography, Disciplinarity, and the Animal Trace," Etienne Benson examines the very process of writing about animals. Benson argues that writing about animals is not just a human endeavor—the activity of writing is coproduced by the animals in our narratives. He thus questions the human assumption of complete control over how animals are represented, arguing instead that we continually interact with interdependent embodied traces. Most of these traces are nonhuman in origin, and all of them are contingent on one another for the construction of meaning. For example, the hunter and the hunted are bound in a relation of mutual constitution and obligation so that predators are shaped and reshaped by their prey. Benson also offers methods for the historian-hunter who seeks to find animal traces in the archives in which they forage. He argues that new practices and data sets drawn from the sciences and humanities are required "to tell a multivocal, multiperspective story in which the voices and perspectives are not exclusively human." By applying this approach, scholars extend the range of historical actors to include the animals and the traces they create.

All of this means that animals have agency, and this theme is taken up in a number of the essays. For example, Benson's call to move beyond seeing animals merely as part of human histories and instead recognize them as subjects is well illustrated by Stacy Rule's "Animal Meaning in T. S. Eliot's *Old Possum's Book of Practical Cats*." Rule's research documents that Eliot's poem does not represent animals in human metaphors, as is commonly argued about his work. She focuses on Eliot's use of three different names for cats to mark their feline individuality—an everyday name, a particular and individual name, and a secret name that is hidden to everyone but the cat itself and one that represents the cat's private self, inaccessible to humans. The third, secret name for cats in Eliot's poetry ensures that the cats about whom he writes exist apart from humans rather than simply being dominated by them. Rule concludes that Eliot's way of writing about the interaction between humans and other animals allows us to reflect on the reality that animals have unique and often unknowable perspectives.

In Benjamin Arbel's essay, "The Renaissance Transformation of Animal Meaning: From Petrarch to Montaigne" our gaze is returned to human perspectives of other animals. Arbel argues that the commonly held assumption that the Renaissance was a time of widespread degradation of animals is unjustified. He finds that Renaissance writers (both leading humanists and others) wrote poems, eulogies, essays, dialogues, and letters that showed a new attitude toward animals that was not present in the Middle Ages. These new attitudes of the Renaissance connected ethical concern with the dynamic distinction between human and animal that pervaded classical culture. The renewed sensitivity toward animals in Renaissance writing shared some common characteristics, including an increased recognition of the individuality of animals, a growing appreciation of animals' mental capacities, and increased moral concern for animals' well-being. By exploring the works of Plutarch and others, Arbel demonstrates that the change in sensitivity developed from Renaissance humanism, with its focus on ethics and secularism and a rediscovery of classical culture. Like Eliot's poetic discussion of the world of cats, poets from ancient times and the Renaissance creatively used language to capture the complex lives of companion animals. Several of these writings used humor interwoven with deep candid feelings of affection.

While Arbel documents evidence of the shifting status of animals in the writings of Renaissance intellectuals, including writings about the passing of their "pets," in her chapter, "Animal Deaths and the Written Record of History: The Politics of Pet Obituaries," Jane Desmond finds similar shifts in the status of companion animals as recorded in contemporary newspaper obituaries. Noting that pet obituaries commemorate a life while ensuring a historical trace of that life in the public record, Desmond finds that these narratives serve as evidence of the enhanced position of the "pet" as a human companion. Because paid obituaries are becoming commonplace, memorializing a treasured companion is now democratized, at least for those with the ability to pay. Paid obits are written by the human family of the deceased pet, not by the newspaper staff, assuring that the obituary makes it into print. Desmond argues that shifting the "boundaries of the includable" is a new phenomenon in obituary writing. The shift is now possible with the declining authority of newspaper editors, who could decide not only if a pet should get an obituary but also whether gay and lesbian partners or unborn children should be commemorated. The critics of pet obits make clear a social ranking based on value. They argue that memorializing a beloved pet is not only ridiculous and in bad taste but also "feminized as excessively mawkish and emotional." Some obits, such as those for hamsters, are more easily parodied than those for a golden retriever or a cat who has been with a family for twenty years. And speaking of family, pet obits join single-parent households and families of gays and lesbians in challenging the very concept of "family." Desmond concludes that pet obits not only highlight a parity between the value of humans and other animals but also could be a harbinger of major shifts in social practices, including a sweeping reorganization of how humans use animals for food, sport, and scientific research.

Sharon Wilcox Adams's "On the Trail of the Devil Cat: Hunting for the Jaguar in the United States and Mexico" focuses on the use of animals for sport and the complexity of agency, a power that can be gained and lost, in part or entirely, with or without intent. Adams examines the narratives of nineteenth- and early-twentieth-century naturalists and hunters to document how the jaguar was encountered and represented. Her research finds that this process of representation left the jaguar removed from her animal-self, taking a new form based on human ideas of "jaguar-ness." Since the documents she examines were marked by class, prestige, whiteness, and masculinity, Adams warns that the production of the jaguar narratives necessarily reflects silences, including that of the jaguars themselves. One of the most problematic aspects of this complex and contested discourse is the animal's relative scarcity and absence, which disproportionately drives certain misconceptions, such as the myth of ferocity that was a popular but unfounded notion for over 100 years.

Unlike the brief intimacy of the hunt, the relationship between a human and his or her assistance dog is an enduring, close, working relationship. Avigdor Edminster extends the significance that Donna Haraway and others attribute to trainers and hunters as bearers of animal traces in his essay, "Interspecies Families, Freelance Dogs, and Personhood: Saved Lives and Being One at an Assistance Dog Agency." Based on observations at a North American assistance dog agency, Edminster finds that the agency's mission to "create relationships" is realized in the enactment and shaping of notions about what humans and dogs can share on multiple levels—physical, emotional, and cognitive. The close bond between clients and assistance dogs allows for a special intimacy between dogs and humans, and their relationships are often understood in terms of family or work (a dog is like a son or daughter, or the dog advertises himself or herself). However, not all human-dog interactions are positively conveyed at the agency. Sometimes dogs are objectified and referred to as inferior to humans (let them know who is boss or bring the dog down like a steer). Edminster discusses assistance dogs who come from kill-shelters as particularly vulnerable in given a last chance to live by proving themselves worthy of assistance work. Indeed, one client who was matched with a shelter survivor felt that she, too, was given a chance to live by the assistance dog. Edminster emphasizes themes of agency, interdependency, relational connections, and personhood. Whether willing or willed, the human-dog partnerships that Edminster analyzes challenge us to reconsider key notions about the relationship between humans and other animals.

In "Mobility and the Making of Animal Meaning: The Kinetics of 'Vermin' and 'Wildlife' in Southern Africa," Clapperton Chakanetsa Mavhunga reveals the signs of "the animal" in southern Africa. His exploration begins by highlighting the subjective status of the label "human" that is placed upon and removed from (some) of us by others. Mavhunga centers the role of the animal as "first author in its designation as an other" and emphasizes the power of the animal by advocating the expansion of Michel Foucault's notion of biopower. By focusing on mobility across borders between human and animal, civilization and nature, wild and domestic, Mavhunga captures the ephemeral and permeable character of these categorizations. He argues that nature is not only inexorably linked to specific historical contexts but is also marked by good animals (wildlife) and bad animals (vermin). Noting that scholars have neglected the study of vermin or pests, Mavhunga's study of the pests and pathogens of Limpopo, southern Africa, extends animal studies scholarship into the realm of African studies and environmental history, a significant disciplinary intersection.

The analysis of the problem animal or pest and contemporary uses of these categories in racist discourse makes a significant contribution to our understanding of ourselves through the othering of others. This intersection is emphasized in Meisha Rosenberg's "Golden Retrievers Are White, Pit Bulls Are Black, and Chihuahuas Are Hispanic: Representations of Breeds of Dog and Issues of Race in Popular Culture." Rosenberg recognizes the profound role of the media in the idolizing of some breeds and the demonizing of others, while powerfully portraying how such representations intertwine with racial and socioeconomic prejudice. She argues that certain dog breeds reflect contemporary society's perceptions of and problems with difference. She analyzes how middle-class whiteness is mapped onto golden retrievers, how pit bulls are identified with lower-class African Americans, and how Chihuahuas are stereotyped as Latinos. Rosenberg paints a provocative portrait of the anxieties around purity and pedigree that characterize the late twentieth and early twenty-first centuries. Analysis of such racializing of the animal expands scholarship concerning pedigrees and pets in the nineteenth century while also enriching our understanding of the application of anthropomorphism as we attempt to map meaning, here in the form of stereotyping, onto the bodies of others.

The questions "What is human?" and "What is animal?" and exploration of how these categories have been, and continue to be, patrolled resonate in both Rosenberg's and Mavhunga's work. The relationship between human and animal is also explored in Analía Villagra's "Cannibalism, Consumption, and Kinship in Animal Studies." Villagra addresses the theme of eating animals by conceptualizing consumption as an act of kinship, arguing that such a relationship can be an empowering experience for both the consumer and the consumed. In so doing, she engages with questions of animal agency and the human-as-animal while challenging us to confront a tension that lies at the heart of animal studies: is it possible to reconcile a sense of kinship with other animals with the practice of consuming their bodies? Villagra argues yes, and she introduces the ultimate villainized human/animal—the cannibal—as a means of reconciling the apparently irreconcilable. By embracing the fluidity that characterizes the status of other animals (at times prey, predator, food, consumer), Villagra argues that eating animal flesh is a process of interspecies kinship. Indeed, she notes "becoming cannibal is not so tragic"; on the contrary, it is an opportunity for a new relation with our animal kin.

Questions of kinship and ethical relations explored in Villagra's work are illustrated in Casey R. Riffel's "Animals at the End of the World: Notes toward a Transspecies Eschatology." In a comprehensive analysis of Philip K. Dick's *Do Androids Dream of Electric Sheep?* Riffel uses apocalyptic time to demonstrate how our understandings of other animals must move beyond history, materiality, and ontology toward new social, ethical, and kin relationships. Riffel's essay is well placed as the final chapter in the volume—a look to the future. The process of looking ahead allows us to encounter animals with new conceptions and ethics, creating human-animal relationships riddled with complexities anew.

Anxieties accompany these complexities. Both Riffel and Villagra challenge us to face our uneasiness about the meanings humans map onto our kin. Through her discussion of cannibalism, Villagra asks, Why should the sudden personhood of an animal discomfit us so, while Riffel's reading of *Do Androids Dream of Electric Sheep* pushes us to break the comfortable confines of ontology. Similarly, Mavhunga and Rosenberg provoke us to recognize the role of the "human" and the "animal" in racial stereotyping, thus highlighting the central place anxiety occupies in contemporary culture. This anxiety has been an undercurrent in much animal studies scholarship. These chapters confront and clarify the origins and meaning of this response to the human-animal relationship.

In many ways, *Making Animal Meaning* ends like it begins, by questioning how animal meaning is made, by whom, and with what consequences. By simultaneously embracing traditional animal studies foci, such as animal agency, identity, and animalization, while emphasizing the extent to which animals can be revealed as authors of meaning, this volume both contributes to and significantly extends animal studies scholarship. Chapters such as those by Benson, Mavhunga, and Villagra reveal new methods for the scholarly endeavor of discovering animal traces and demonstrate the ways in which interdisciplinary and international perspectives allow for the centering of the animal actor. We see how complex human-animal relationships can be revealed through a broad spectrum of sources, including the historical illustrations and texts of jaguar hunters (Adams) and the contemporary lived relationships between humans and the animals who assist them (Edminster). Ranging from the philosophers of the Renaissance (Arbel) to the poets of the nineteenth century (Rule), to the obituary columns (Desmond) and popular culture of the present (Rosenberg), to a fictional timeless world (Riffel), the chapters eloquently demonstrate the significance of animal meaning for our distant and more recent past, our present, and our future.

PART 1

Making New Animal Meanings

Animal Writes

Historiography, Disciplinarity, and the Animal Trace

ETIENNE BENSON

THOSE OF US WHO ATTEMPT TO WRITE ABOUT NONHUMAN ANIMALS ARE ALL IMPLI-
cated by the pun that appears in the title and throughout the text of Jacques Derrida's *L'Animal que donc je suis*.[1] I follow or track (*suis*, from the infinitive *suivre*) the animals about whom or about which I write, and I also am (*suis*, from *être*) an animal—specifically, a writing animal. This doubleness of animal writing—its way of situating us simultaneously as subject and objcct, autobiographer and biographer, pursued and pursuer—is evocatively captured in the opening line of Philip Armstrong's study, *What Animals Mean in the Fiction of Modernity*, as is the powerful and pervasive assumption that writing is a uniquely human activity: "An animal sits at a desk, writing."[2] To which we could add, "writing about animals," which is always a pursuit both of the other and of ourselves: the animal that I follow, the animal that I am.

Like any pursuit, writing about animals depends on tracks, trails, or traces—those material-semiotic remnants of whatever it is the pursuer hopes to catch, those often unintentional indexes of a now-absent presence.[3] In this essay, I consider the relationship that is established in the course of writing, where "writing" is understood as a form of tracking and leaving tracks that is less specific to the human species than is usually assumed. I am especially concerned with the difficult case of historical writing, in which the "real" animal in whose gaze or body the author and art critic John Berger and many others have hoped to root an authentic, genuine, or ethical relationship has long since vanished.[4] In the absence of the living animal body—in the presence of the dead, one might say—the animal historian must instead forge a (real, genuine, authentic, ethical) relationship with the embodied traces of past animal life. Doing so, I argue, requires jettisoning some of the assumptions about historiographic practice and disciplinary identity that have heretofore largely defined the scholarly field of animal studies and the narrower subdiscipline of historical writing about animals.

One of those assumptions concerns the nature of the sources from which the historian reconstructs past animal lives, whether human or otherwise. To the extent that historical sources are understood in conventional terms, that is, as textual or linguistic documents or records, the dilemma is clear. Such sources can provide rich descriptions and important insights into historical changes in human attitudes toward and relationships with animals, as the growing literature of animal history amply demonstrates.[5] But they suffer a profound limitation from the perspective of the historian who wants to tell a multivocal, multiperspectival story in which the voices and perspectives are not exclusively human. Textual sources seem always to arise from the experience or activity of one particular kind of animal—the writing animal, the human. This is true even when, and perhaps particularly when, they claim to speak in the voice or see from the perspective of nonhuman animals. We might, as conservationist Carl Safina expresses it in his book *Eye of the Albatross*, want to use all of our human skills and resources "to draw out what the animals cannot tell us," to "give words to the wordless, and voice to the voiceless."[6] But in doing so we are always at risk, as Donna Haraway reminds us, of ventriloquism—of speaking for rather than allowing to speak, of talking before we listen.[7]

Erica Fudge has argued that because human historians only have access to animal lives through human documents, a "history of animals . . . is impossible," strictly speaking. To the extent we continue to use the phrase, it must be *sous rature*—crossed out but still legible—in recognition of the fact that such a practice is impossible even as the desire for it shapes what we do. Instead, we must be satisfied, she suggests, with a history of "human attitudes toward animals," one that is, at its best, sensitive to the way "the human" and "the animal" have been co-constructed in theory and related in practice—that is, a "holistic" history combining attention to both discursive and real animals.[8] My approach to historical sources and the possibility of animal history is somewhat different. The apparent impossibility of animal history is, I believe, precisely a result of adopting a strict, hierarchical division between the real and the discursive, things and representations, animals and attitudes toward animals. It becomes less compelling when those divisions are suspended.

As Haraway writes in *Primate Visions*, by way of explanation of her use of the term "material-semiotic," nonhuman animals "act and signify, and like all action and signification, theirs yield no unique, univocal, unconstructed 'facts' waiting to be collected. . . . Like words, machines, equations, institutions, generic writing conventions, people, and landscapes, the animals have specific kinds of solidity in the apparatus of bodily production."[9] I take this to have a twofold meaning: first, that animals have a "solidity" or presence in written documents, scientific or otherwise, that goes beyond mere "representation"; second, that humans are not the only creatures who leave meaningful marks or traces on the body of the world (which has in turned marked them).[10] A certain recent turn in scholarly understanding of human-animal relationships and the human-animal distinction makes it clear that to see even the most human of texts as nothing *but* human—as entirely devoid of particular animals, "the animal" in general, or animality—is to succumb to one of the most powerful modern humanist illusions. If we have never been modern, as Bruno Latour suggests, neither have we (or the traces we leave behind) been human, if "human" is meant to indicate a class of beings separated from all other living beings by an unbridgeable ontological or ethical abyss.[11] Everything we do, including writing, is shaped by our long evolutionary history of interactions with other animals and our present lived interdependence with them.

That this is true has been perhaps most persuasively argued in studies of hunting and domestication. In books such as *The Tender Carnivore and the Sacred Game* and *The Others*, primitivist Paul Shepard argues that humanity discovered itself through embodied relations with wild animals, particularly those relations established through hunting.[12] Studies of modern-day indigenous hunters, such as Hugh Brody's studies of hunters in the Arctic, Tom Ingold's studies of Saami reindeer herders, Louis Liebenburg's study of South African hunters, or Rane Willerslev's study of the Yukaghir hunters of Siberia, have shown that the most successful hunters are adept at imagining themselves into the minds and bodies of the animals they seek to kill in order to understand, predict, and manipulate their behavior.[13] This obligatory act of imagination, Willerslev argues, puts the hunter in the "paradoxical position of mutual mimicry," a position whose effects extend well beyond the practice of hunting to shape the hunters' social relations of all kinds (that is, its relations with humans as well as with nonhuman animals).[14]

Even scholars who do not privilege hunting or share Shepard's belief that the disappearance of bloody worship from the lives of most modern humans has made us inhuman—a belief that differs from Berger's particular form of Marxist humanism mainly in its location of the prelapsarian among hunter-gatherers rather than among precapitalist peasants—have agreed with the basic premise that animals and animality are constitutive of the human. Take studies of domestication, which were once founded on the assumption that domestication was an example—perhaps the example par excellence—of human domination over nonhuman animals.[15] That view has increasingly given way to the perspective popularized by Stephen Budiansky in his books *The Covenant of the Wild* and *The Truth about Dogs*, in which the central question is not how humans reshaped animals for their own purposes but "why animals chose domestication."[16] In so choosing, the argument goes, nonhuman animals—the wolves who may have scavenged at the edge of human settlements or partnered with human hunters; the wildfowl who submitted themselves to the eventual ax in exchange for food, shelter, and a dramatic expansion of reproductive possibilities; and so forth—reshaped themselves for their own goals, while also profoundly reshaping human societies and cultures. That coevolutionary process continues today. Human society is not merely built on the backs of nonhuman animals; it is also built for and by animals.[17]

This line of thought can be taken to absurd extremes. The intellectual and ethical contortions required to see the industrial production of low-cost chicken products as a mechanism of avian flourishing are not worth the effort, not only and not least because they require a conflation of the welfare of the individual and collective (or, more precisely, an erasure of the individual in favor of the collective) and a reduction of flourishing to mere numerical abundance and biogeographical distribution.[18] We should hesitate to wish such a form of flourishing even on our enemies. But a more modest version of the thesis does obtain. As a result, to a limited but important extent, writing about human history is always-already writing about animals, regardless of whether the writer has any interest in or knowledge of dogs and chickens. Humans are a kind of animal that (like all kinds of animals) has been and continues to be profoundly reshaped by its interactions with other kinds of animals.[19] Even writing and reading, those seemingly quintessential human activities, can be seen as having roots in animal signs, hunting, and tracking, as J. Edward Chamberlin has argued.[20] All history is animal history, in a sense—that is, history written by, for, and about animals. The only question is which.

But of course there remains a difference between historical scholarship whose explicit focus is limited to the human animal, even if that includes the impact of other animal life on humans and recognizing the animality of (human) writing as an embodied practice, and historical scholarship that aims in some way to explore the history of nonhuman animals as subjects in their own right and for their own sakes. For this latter kind of history, human-authored texts can still provide valuable insights into the past that are not reducible to the human perspective. The same operation of intellectual judo that has made it possible to see domestication more as a partnership, however unequal, than as a simple case of domination or objectification can also make human-authored texts about animals seem more like the result of a collaboration or coauthorship—a collection of traces of the animal who writes through the human as well as of the human who writes about the animal.

That such traces cannot be reduced to their human authors is perhaps most evident when one moves away from strictly linguistic textual records toward other sorts of "animal traces," such as the early wildlife photographs whose history Matthew Brower has recounted, in which animals are present in ways that clearly exceed the intentionality of their human authors.[21] It would be mistaken, however, to see only such pictorial, photographic, cinematic, or sculptural (for example, taxidermic) productions, which have an iconic as well as indexical relationship to the animals they depict, as being coproduced by their subjects. Because writing in any existential mode is a practice embedded in a real world of which nonhuman animals are an integral part, even avowedly fictional texts may contain traces of nonhuman animals. Pamela Banting has suggested that certain practices of attentive writing—she focuses on several late-twentieth-century Canadian writers who write about nature and wild animal life—are particularly amenable to revealing the trace of the animal.[22] Attentive reading may also bring to light the traces of nonhuman animals even where their (human) authors did not intend to make them visible. John Simons, in his *Animal Rights and the Politics of Literary Representation*, suggests that "my apprehension and analysis of what is going on in a cultural text and specifically a piece of literature should be quite different if I read it not for the signs and traces of human struggle but rather for the tracks of the animals with which we share the planet."[23] Human writing in a world where human life is so intricately intertwined with nonhuman life will inevitably reveal the traces of the other.

In practice, however, even if it is both possible and critically important to look for the traces of the animal, animals, or animality in even the most human of texts, the challenges for animal history can seem of a different order than those for the history of humans who have not left textual records. One may use techniques to look for the traces of nonhuman animals in human archives that are superficially similar to those one uses to look for the traces of subaltern, poor, or disenfranchised humans in the archives of the powerful, but there seems to be a deeper divide.[24] As Fudge has noted, the project of animal history raises questions about the necessity of intentionality and temporal consciousness (or, as she puts it, a "concept of historical periodization") for history.[25] Even when sources about past animal life are abundant, animal history faces an ontological challenge that can be expressed as a variant of Wittgenstein's much-cited hypothesis that if a lion could speak we would not understand him; namely, that if lions could leave historical records they would not reveal a history that we understand as such—that is, a progressive, linear history in which the essence of what it is to be a lion depends on where and when a particular lion finds himself or herself.[26]

The kind of agency required for such a progressive history—which remains history proper, despite efforts to produce "nonlinear" or deconstructionist histories[27]—seems to require something more than offered by relational theories of agency such as actor-network theory, in which the agency of any particular entity arises from its network of relations to other entities. When Timothy Mitchell asks whether the mosquito can speak, he means to ask whether nonhuman entities can exceed and confound human intentions, the answer to which must surely be affirmative. But that mosquitoes themselves might have a history worth telling remains beyond the pale.[28] This widely shared if rarely explicitly stated hypothesis rests on two assumptions. The first is a theoretical claim about the minimal properties required of an entity before it can be considered a legitimate historical subject or actor, which usually center on intentionality, language, and consciousness of the past and future, including the consciousness of death. (These properties may be assumed to be held in general or by the collective, rather than by the individual.) The second assumption is a factual one, sometimes supported by casual reference to the scientific literature—namely, that these are properties held by humans but not by other animals.[29]

My argument here rests on two interrelated counterassumptions. First, whether animals are properly historical subjects or actors is an open, empirical question, not one that can be answered from the armchair. The answer may vary—is highly likely to vary, I would argue—according to the particular historical situation and kind of animal in question. To assert that animals have no history, properly speaking, is to assert a transhistorical truth, an operation perhaps appropriate for metaphysicians but not for historians committed to a rigorous historicism. It does not matter whether we see the historylessness of animals as a gift, as Friedrich Nietzsche and Rainer Maria Rilke did; as a lack, as Martin Heidegger did; or as a messianic promise, as Giorgio Agamben does. In all of these cases we would be assuming in advance what should be the result of our research.[30] Second, the answer to this question and the methods that are necessary to answer it are within the professional domain of the historian. Though historians might need to draw on natural science and other fields outside the traditional purview of history—and in fact I argue below that they must—scientists no more than philosophers can tell them whether it is possible to tell the history of a particular subject. It is for historians to decide what is properly historical.

In doing so we would be better served by broadening our conception of what counts as a legitimate "primary source" and working out the practical consequences of such a broadening than we would be by quasi-philosophical disputation over the nature of animal agency, voice, or perspective, no matter whether such disputation is conducted on the grounds of actor-network theory, poststructuralist feminism, or other theoretical frameworks. Sufficient disputation has taken place to make it clear that we should at least entertain the possibility that nonhuman animals may be approached as legitimate historical actors.

Although several historical studies in recent years have already begun testing the boundaries of this possibility, these studies remain limited in the range of sources that they use and in their deployment and historicization of natural-scientific data and insights into animal behavior.[31] Though it is a fair beginning, I do not think we should be satisfied by the use of today's scientific understanding of wolf behavior, for example, as a transhistorical constant that can be projected back onto the historical and ecological conditions of two centuries ago, much as an anthropologist might project social structures and processes identified in present-day hunter-gatherer societies back onto the earliest human societies.[32] Whether such an operation is valid—that is, whether a particular kind of nonhuman animal (or a certain group of "primitive" or "noncivilized" humans) is a properly historical subject whose essence is contingent on the time and place of its being—is exactly what we should hope to find out in the course of our research. In other words, having admitted the theoretical possibility of nonhuman animal historical agency, we should pose the actual existence of such agency as an empirical question rather than a philosophical or ontological presupposition.

Where possible, that research should, as I have suggested above, go beyond the important work of reading the traces that nonhuman animals have left in human texts. In *The Spell of the Sensuous*, David Abram argues for an expanded understanding of signification in which the traces of nonhuman activity with which the world is marked are seen as different perhaps in degree and form but not in kind from the linguistic traces of literate humans. Drawing on Maurice Merleau-Ponty's account of embodied perception, Abram argues that "linguistic meaning is primarily expressive, gestural, and poetic, and that conventional denotative meanings are inherently secondary and derivative." If one accepts that premise, one must renounce,

Abram argues, "the claim that 'language' is an exclusively human property. If language is always, in its depths, physically and sensorially resonant, then it can never be definitively separated from the evident expressiveness of bird-song, or the evocative howl of a wolf late at night. . . . Language as a bodily phenomenon accrues to all expressive bodies, not just to the human."[33] Although Abram is not primarily concerned with temporality or historicity, his account can easily be extended in that direction. Expressive bodies often leave more durable traces than wolves' howls; and even a howl can, when traced in the right medium, survive to be interpreted and reinterpreted long after the death of the howler.[34]

Abram poses his account of natural signification against the foil of poststructuralism, particularly Derrida's account of the world as text, which Abram sees as ignoring the vital role of nonhuman entities in constructing a common world.[35] However, as Cary Wolfe points out in *Animal Rites*, the deconstructionist slogan that there is nothing beyond the text is not to be read as an idealist claim about the construction of reality through human writing, language, or thought. On the contrary, it has more to do with the materialist second-order or reflexive cybernetics of Humberto Maturana and Francisco Varela than it does with any sort of textual reductionism, and thus makes a move very similar, albeit from a different direction, to the one made by Abram. Where Abram seeks to identify the sensual and physical aspects of language, the deconstructionist project seeks to identify the linguistic aspects of the sensual and physical world. Both projects meet in what might be called the material-semiotic middle—that is, the recognition that meaning is embodied and that all bodies participate in the making of meaning.[36] For both, an expansive and ahumanist notion of "writing" is central.

The relevance for rethinking the primary source base upon which we write animal histories should be clear, even if the practical implications remain to be worked out. As Banting argues, "including animal calls, tracks and other forms of bio-, zoo- or eco-semiotics in our notion of text" opens up the possibility of rethinking "our notions of signature, event and habitat, as well as those of subjectivity, voice, writing, author and authority, setting, metaphor, and meaning."[37] This is true not just for the present-day relations to animals upon which Banting focuses, but also for our relationship as writers and historians to the embodied traces of past animal life. The material-semiotic world is full of traces of the past, some of them human in origin, the vast majority not. All of these traces are meaningful only in relation to other traces, whose meaning, in turn, is incompletely and unstably determined by their own sets of relations. The world is a fleshly text—an embodied collection of interdependent traces.[38]

Geoffrey Bowker has pointed out that the idea of Earth as an archive was a common metaphor in the writings of mid-nineteenth-century geologists, who sought to reconstruct the deep past—a project in which a certain kind of animal trace, fossils, played a central role, as of course they also did in that other great nineteenth-century historical project, the development of evolutionary theory.[39] My suggestion here is that we take this metaphor and its kin seriously, and that we attempt to transcend the disciplinary divide that has given the interpretation of animal traces over to the natural sciences while reserving the interpretation of human traces to the social sciences and humanities. This does not have to—and I would argue should not—entail importing scientific (or scientist) models of animal behavior, evolution, or ecology into the humanities. As Edmund Russell has argued in his prospectus for "evolutionary history," attention to the biological or ecological need not entail the wholesale adoption of sociobiological or evolutionary-psychological interpretive frameworks, whether applied to humans or nonhuman animals. Nor need it entail focusing exclusively on humans, as Russell suggests by limiting his focus to "the role of evolution in human history." My hope would be for something broader, something beyond the human—an approach in which the term "history" need not imply humanity any more than the term "evolutionary" does.[40]

Reconceptualizing historical sources in terms of material-semiotic traces of the past is, I believe, an important first step in opening up the possibility of such a field. Contra Fudge, it is simply not true that the "only documents available to the historian in any field are documents written, or spoken, by humans," unless one takes a question-beggingly narrow view of what counts as a "document"—a view that seems inconsistent with the idea that documents may be "spoken."[41] Even within relatively conventional understandings of historical method, there is a venerable tradition of challenging such narrow understandings of the range of legitimate sources or documents for historical study.

Marc Bloch, the medieval historian and cofounder of the Annales school, argued in his posthumously published methodological meditation, *The Historian's Craft*, that the historian's knowledge of past events or actors was always "knowledge of their tracks."[42] Like François Simiand, the historical economist from whom he borrowed the phrase, Bloch sought to reconstruct the past lives of people—medieval peasants, for example—who had left little in the way of historical documents as they were conventionally defined. The trope of traces broadened the realm of potential sources from written texts to the entire world, so long as that world could be interpreted as in some way referring to the past. It also emphasized the materiality of the world of signs, marks, and remnants—the sensible, tangible things through which the historian encountered the past. As Bloch put it: "Whether it is the bones immured in the Syrian fortifications, a word whose form or use reveals a custom, a narrative written by the witness of some scene, ancient or modern, what do we really mean by *document*, if it is not a 'track,' as it were—the mark, perceptible to the senses, which some phenomenon, in itself inaccessible, has left behind?"[43] New kinds of historical documents would bring new kinds of historical subjects—those phenomena in themselves impossible to grasp—into the historian's purview.

Historiographical theorists have also challenged the idea that historical interpretation is limited to the interpretation of the explicit, intentional, surface themes—the denotative content—of written documents. In his essay "Clues: Roots of an Evidential Paradigm," microhistorian Carlo Ginzburg uses tracking as the core metaphor for an understanding of historical methodology.[44] Historical scholarship, Ginzburg argues, is among the "divinatory" or "conjectural" modes of knowledge production—a category in which he includes art criticism, criminal forensics, psychoanalysis, medical diagnostics, and, originally and most fundamentally, the tracking of wild animals by human hunters. These disciplines operate according to the assumption that "infinitesimal traces permit the comprehension of a deeper, otherwise unattainable reality."[45] These traces are often revealing exactly because they are unintentional; the rabbit runs in order to escape, not in order to leave tracks in the snow. This divinatory style of interpretation, Ginzburg suggests, originated with primitive hunting, as did narrative itself. Just as a physician draws conclusions about a patient's health on the basis of subtle changes in the rhythm of the heart or the sound of the breath—or, these days, from a protein marker or genetic test—so a hunter infers the past activities and present location of prey from a bent twig or a partial paw print. "The hunter would have been the first 'to tell a story,'" Ginzburg writes, "because he alone was able to read, in the silent, nearly imperceptible tracks left by his prey, a coherent sequence of events."[46] The historian's hunt for the past similarly relies on his or her ability to imaginatively reconstruct coherent sequences of events—and their underlying causes or deeper meanings—from fragmentary, inconclusive traces.

Although Bloch's expansive notion of historical sources and Ginzburg's comparison between the tracking of animals and the writing of history would seem to open the way toward including animals as historical actors or agents, neither of these authors took that leap. Bloch was quite explicit that only traces of human life should be of interest to the historian—that it was, in fact, the unique complexity of human behavior that required the historian to draw on a wide range of textual and nontextual sources. Those traces might well include the traces left by animals, but they were only properly "historical" to the extent that they shed light on human lives (much as Russell's "evolutionary history" is distinguished from the evolutionary history of biologists by its focus on anthropogenic evolution).[47] Ginzburg's argument, which relies on an implicit historical progression from hunting to history, is equally anthropocentric. Though animals are clearly vital actors in the narratives told by primitive hunters, they vanish from the analogous modern-day practices of physicians, art critics, historians, and so forth, which are Ginzburg's true subject. Animals and hunting play an originary role, providing the stuff of stories in the dawn of time; today, his account implies, they have been superseded by more civilized concerns linked to the extension of modern apparatuses of social control.[48] Animals were, in a sense, the training ground on which humans learned the techniques and ways of thought that would allow them to track and control other humans.

Still, the tropes of traces and tracking are amenable to less anthropocentric interpretations than these. They provide a way of thinking about historical sources that does not depend on the assumption that the producer of a trace was, in a particular instance or in general, intending to leave a record or to convey a specific meaning. The vast majority of historical sources produced by humans are not, of course, intended for the eyes of historians; in many cases their authors would be surprised at the uses to which they have been put. That they are useful for a particular historical study is a by-product or accident of the fact that they were once useful for other purposes and have since been preserved and made accessible. Indeed, as Ginzburg suggests, the most valuable clues to the past are often exactly those traces that their producers did not intend to leave.[49] Moreover, as Derrida has argued, there is much that is automatic in even the most conscious linguistic act of a self-aware human. The distinction between an automatic, unintentional, meaningless animal "reaction" and a willed, intentional, meaningful human "response," which has been a core tenet of philosophical humanism since at least the time of René Descartes, rests on shaky empirical and philosophical ground.[50] If intentionality were to serve as a litmus test for the legitimacy of historical sources or agents, much of human history would have to be thrown out. In the same way, the traces left by other kinds of animals besides the human need not have been left intentionally to be useful for the historian. Whether the traces make it possible to tell an interesting or meaningful story about nonhuman animals as historical actors is another question—as suggested above, an empirical one, the answer to which is likely to vary, just as it does for the history of humans, depending on the available sources, the historical situation, and the historian's interests and capacities.

As Ginzburg suggests, there is a sense in which the historian's method resembles that of the hunter. It seeks to imaginatively reconstruct a sequence of past events on the basis of often fragmentary evidence. While the successful hunter returns home with the body of the prey, the successful historian returns with a plausible and interesting explanation of change over time—or, in its minimal form, the bare demonstration that a particular past time was different in some nontrivial way from the present. Success depends not just on the talents and resources of the hunter or historian but also on the conditions under which he or she works. The most skilled tracker may lose the trail on bare rock; the most diligent and perceptive historian will have little to say when time has effaced the traces of the past.

The current paucity of traces for the hunter-historian to follow is neither accidental nor innocent. It is a product of the very history we want to tell. The production, preservation, and availability of historical traces is often a costly effort that reflects existing power relations and assumptions about the value of traces and their makers; the range of available sources at any given time is a product of the entire intervening history of such values. The embodied traces of the past survive to meet the historian's eye (or ear or hand or nose) only because of a long, often circuitous and richly textured history—a history that begins before the trace is even produced, that establishes the trace's conditions of possibility, and that therefore cannot be disentangled from whatever historical phenomenon it is we want to study or from our own historical situations. Writing requires a marker and a medium—pen and paper, figuratively speaking—and both are historically contingent, whether the writer in question is a human or another sort of animal. The marks that a particular cottontail rabbit and the owl who pursued him or her left in the winter snow more than a century ago would not have been left in the summer grass (see figure 1). Nor would those traces have survived—much altered by their transduction into another medium, but not entirely betrayed—to meet the early-twenty-first-century historian's eye unless they had been traced in the sketchbook of a naturalist and published in a popular magazine, of which a few carefully preserved copies survive in libraries.[51]

In the past several centuries the varieties of figurative pens and paper with which animals can leave traces for historians to read has multiplied and diversified to a remarkable extent. Since the late 1980s, to take a recent example, it has been possible for properly equipped birds to leave hourly traces of their long-range migrations in the data banks and on the computer screens of the scientists who have attached satellite-linked radio tags to them. Many thousands of other individual birds leave traces of their presence when they are banded and recaptured or resighted by professional and amateur birdwatchers.[52] These traces are not merely human "representations" of animal movements or behavior, any more than the transcript of a court proceeding or a personal diary is merely a "representation" of the witnesses' speech or the diarist's thoughts. In all of these cases, which of course differ in many other ways, meaning is coproduced within a technological system of trace-making, a network of material-semiotic articulations.[53] It would be as mistaken to dismiss the diary as providing no insight to the thoughts and feelings of the diarist as it would be to take it as a transparent window into the soul. Similarly, technical systems that make it possible for nonhuman animals to leave new kinds of traces on the world, even if built by humans, cannot be reduced to human intentionality.

Needless to say, documenting the lives of animals for their own sakes has heretofore not been a high priority of archivists, historians, or governments. Still, many potential archives are made, collected, and preserved for purposes very different than those to which later historians put them. The explosion of documentation with which modern historians are familiar, and which is tied to the state and other apparatuses of social control noted by Ginzburg, is not limited to the human; nor is it a case of extension to animals of techniques originally designed for humans, or vice versa. Both have developed as part of a single differentiated system.[54] Scientific research is one potential source of such protoarchives, as are the records of industries and government agencies that deal closely with nonhuman animals in one way or another—whether as pests, as pets, or as sources of profit. With their noninnocent conditions of production kept firmly in view, such sources may provide the basis for a richer history—a true "animal history," rather than merely a history of "human attitudes toward animals." It is in the nature of archives to be built for futures that may never come, and it is one of the tasks of the historian to reappropriate such archives in the future that does.

Figure 1. Canadian nature writer Ernest Thompson Seton argued that animals were capable of signifying through tracks and other signs, which he described as a kind of protowriting. Such inscriptions revealed a world of animal life that was otherwise invisible. "It is a remarkable fact that there are always more animals about than any but the expert has an idea of," he wrote. Ernest Thompson Seton, "The Oldest of All Writing—Tracks," *Country Life in America* 17 (December 1909): 169. This sketch of a "woodland tragedy" is on 73.

Learning to reappropriate such archives, and perhaps even in some cases produce them, is one of the ways in which disciplinary boundaries of history might usefully be stretched. Historians who are interested in telling stories about the past in which nonhuman animals are central have good reasons to support the development of a data infrastructure that would make it easier to make contact—really, authentically, responsibly—with the embodied traces of the past animal lives. The nature of this archive, down to the concrete details of organization and infrastructure, will shape the kinds of histories that can be written. Just as political historians work to ensure that government documents remain accessible to themselves and to future historians, or social historians gather oral histories to ensure that certain stories are not lost, so, too, might animal historians work to ensure that the embodied traces of animals (that is, all animals, not just the human kind) are preserved, reproduced, and archived so that better stories about animals, human and otherwise, can be told.[55]

NOTES

1. Jacques Derrida, *The Animal That Therefore I Am*, trans. Marie-Louise Mallet (New York: Fordham University Press, 2008).
2. Philip Armstrong, *What Animals Mean in the Fiction of Modernity* (New York: Routledge, 2008), 1. This opening line is in dialog with the image on the cover of Armstrong's book, a painting by Peter Zokosky of a chimpanzee sitting at a desk with pencil, paper, and coffee mug.
3. "Material-semiotic" is drawn from Donna J. Haraway, *Primate Visions: Gender, Race, and Nature in the World of Modern Science* (New York: Routledge, 1989), 310–311.
4. John Berger, "Why Look at Animals?" in *About Looking* (New York: Vintage, 1980), 3–28.
5. A seminal text is Harriet Ritvo, *The Animal Estate: The English and Other Creatures in the Victorian Age* (Cambridge, Mass.: Harvard University Press, 1987).
6. Carl Safina, *Eye of the Albatross: Visions of Hope and Survival*, (New York: Henry Holt, 2002), 98. Safina's book is an inheritor of the tropes and traditions of the realistic wild animal story; see Ralph H. Lutts, *The Nature Fakers: Wildlife, Science and Sentiment* (Charlottesville: University Press of Virginia, 2001); Ralph H. Lutts, "The Wild Animal Story: Animals and Ideas," in *The Wild Animal Story*, ed. Ralph H. Lutts (Philadelphia: Temple University Press, 1998), 1–22.
7. Donna J. Haraway, "The Promises of Monsters: A Regenerative Politics for Inappropriate/d Others," in *The Haraway Reader* (New York: Routledge, 2003), 63–124.
8. Erica Fudge, "A Left-Handed Blow: Writing the History of Animals," in *Representing Animals*, ed. Nigel Rothfels, (Bloomington: Indiana University Press, 2002), 6. Cf. Susan J. Pearson and Mary Weismantel, "Does 'the Animal' Exist? Toward a Theory of Social Life with Animals," in *Beastly Natures: Animals, Humans, and the Study of History*, ed. Dorothee Brantz (Charlottesville: University of Virginia Press, 2010), 17–37.
9. Haraway, *Primate Visions*, 310–311.
10. See also the notion of "cyborg writing" in Donna J. Haraway, "A Manifesto for Cyborgs: Science, Technology, and Socialist Feminism in the 1980s," in *The Haraway Reader*, 33.
11. Bruno Latour, *We Have Never Been Modern* (Cambridge, Mass.: Harvard University Press, 1993).
12. Paul Shepard, *The Tender Carnivore and the Sacred Game* (New York: Scribner, 1973); Paul Shepard, *The Others: How Animals Made Us Human* (Washington, D.C.: Island Press, 1996).
13. Rane Willerslev, *Soul Hunters: Hunting, Animism, and Personhood among the Siberian Yukaghirs* (Berkeley: University of California Press, 2007); Tim Ingold, "On Reindeer and Men," *Man* 9, no. 4 (1974): 523–538; Hugh Brody, *The Other Side of Eden: Hunters, Farmers, and the Shaping of the World* (New York: North Point, 2000); Louis Liebenberg, *The Art of Tracking: The Origin of Science* (Claremont, South Africa: David Philip, 1990).
14. Willerslev, *Soul Hunters*, 11.

15. See, for example, Yi-Fu Tuan, *Dominance and Affection: The Making of Pets* (New Haven, Conn.: Yale University Press, 1984); Marjorie Spiegel, *The Dreaded Comparison: Human and Animal Slavery* (New York: Mirror Books, 1996).

16. Stephen Budiansky, *The Covenant of the Wild: Why Animals Chose Domestication* (New York: W. Morrow, 1992); Stephen Budiansky, *The Truth About Dogs: An Inquiry into the Ancestry, Social Conventions, Mental Habits, and Moral Fiber of Canis Familiaris* (New York: Viking, 2000).

17. For a survey of relevant literature, see Harriet Ritvo, "Animal Planet," *Environmental History* 9, no. 2 (2004): 204–220.

18. On the making of the industrial chicken, see Susan Squier, "Fellow-Feeling," in *Animal Encounters*, ed. Tom Tyler and Manuela Rossini (Boston: Brill, 2009), 173–196; Roger Horowitz, "Making the Chicken of Tomorrow: Reworking Poultry as Commodities and as Creatures, 1945–1990," in *Industrializing Organisms: Introducing Evolutionary History*, ed. Philip Scranton and Susan R. Schrepfer (New York: Routledge, 2003), 215–235.

19. Donna J. Haraway, *The Companion Species Manifesto: Dogs, People, and Significant Otherness* (Chicago: Prickly Paradigm Press, 2003); Vicki Hearne, *Adam's Task: Calling Animals by Name* (New York: Knopf, 1986). For an analogous argument applied to plants, see Michael Pollan, *Botany of Desire: A Plant's Eye View of the World* (New York: Random House, 2001).

20. J. Edward Chamberlin, "Hunting, Tracking and Reading," in *Literacy, Narrative and Culture*, ed. Jens Brockmeier, Min Wang, and David R. Olson (Richmond, Surrey: Curzon Press, 2002), 67–85. See also David Abram, *The Spell of the Sensuous: Perception and Language in a More-Than-Human World* (New York: Vintage Books, 1996); Liebenberg, *Art of Tracking*.

21. Matthew Francis Brower, "Animal Traces: Early North American Wildlife Photography" (Ph.D. diss., University of Rochester, 2005). On animal cinematography and taxidermy, see Hanna Rose Shell, "Things Under Water: Etienne-Jules Marey's Aquarium Laboratory and Cinema's Assembly," in *Making Things Public: Atmospheres of Democracy*, ed. Bruno Latour and Peter Weibel (Cambridge, Mass.: MIT Press, 2005), 326–332; Hanna Rose Shell, "Skin Deep: American Taxidermy, Embodiment and Extinction," in *The Past, Present & Future of Natural History Museums*, ed. A. Leviton (San Francisco: Academy of Sciences, 2004), 88–112.

22. Pamela Banting, "Magic Is Afoot: Hoof Marks, Paw Prints and the Problem of Writing Wildly," in Tyler and Rossini, *Animal Encounters*, 27–44.

23. John Simons, *Animal Rights and the Politics of Literary Representation* (New York: Palgrave, 2002), 5–6.

24. Studies that promise to take animal alterity as seriously as human alterity often fail to follow through in practice; see, for example, Aaron Skabelund, "Can the Subaltern Bark? Imperialism, Civilization, and Canine Cultures in Nineteenth-Century Japan," in *JAPANimals: History and Culture in Japan's Animal Life*, ed. Gregory M. Pflugfelder and Brett L. Walker (Ann Arbor: Center for Japanese Studies, University of Michigan), 195–243.

25. Fudge, "Left-Handed Blow," 6.

26. On Wittgenstein's lion, see Cary Wolfe, "In the Shadow of Wittgenstein's Lion: Language, Ethics, and the Question of the Animal," in *Zoontologies: The Question of the Animal*, ed. Cary Wolfe (Minneapolis: University of Minnesota Press, 2003), 1–58; Simon Glendinning, *On Being with Others: Heidegger, Derrida, Wittgenstein* (London: Routledge, 1998), 70–75.

27. See, for example, Manuel De Landa, *A Thousand Years of Nonlinear History* (New York: Zone Books, 1997).

28. Timothy Mitchell, *Rule of Experts: Egypt, Techno-Politics, Modernity* (Berkeley: University of California Press, 2003), 19–53; see also Linda Nash, "The Agency of Nature or the Nature of Agency?" *Environmental History* 10, no. 1 (2005): 67–69.

29. For an example, see the brief discussion about whether salmon can be "at" the table or merely "on" the table in William Cronon, "Speaking for Salmon," in Joseph E. Taylor III, *Making Salmon: An Environmental History of the Northwest Fisheries Crisis* (Seattle: University of Washington Press, 1999), ix–xi; Frieda Knobloch, review of *Making Salmon: An Environmental History of the Northwest Fisheries Crisis*, by Joseph E. Taylor III, *American Historical Review* 106, no. 4 (October 2001): 1373–1374.

30. Friedrich Nietzsche, "On the Uses and Disadvantages of History for Life," in *Untimely Mediations*, ed. Daniel Breazeale, trans. R. J. Hollingdale (Cambridge: Cambridge University Press, 1997), 57–124; Rainer Maria Rilke, "The Eighth Elegy," in *The Selected Poetry of Rainer Marie Rilke*, ed. and trans. Stephen Mitchell (New York: Vintage, 1989), 193–197; Martin Heidegger, *The Fundamental Concepts of Metaphysics: World, Finitude, Solitude* (Bloomington: Indiana University Press, 1995). See Giorgio Agamben, *The Open: Man and Animal*, trans. Kevin Attell (Stanford, Calif.: Stanford University Press, 2004), 39–74; Matthew Calarco, *Zoographies: The Question of the Animal from Heidegger to Derrida* (New York: Columbia University Press, 2008), 15–54.

31. Virginia DeJohn Anderson, *Creatures of Empire: How Domestic Animals Transformed Early America* (New York: Oxford University Press, 2004).

32. Jon T. Coleman, *Vicious: Wolves and Men in America* (New Haven, Conn.: Yale University Press, 2004); for a human example, see Steven J. Mithen, *Thoughtful Foragers: A Study of Prehistoric Decision Making (*New York: Cambridge University Press, 1990).

33. Abram, *Spell of the Sensuous*, 80; Maurice Merleau-Ponty, *Phenomenology of Perception* (New York: Routledge, 1945).

34. For a less sanguine perspective on the interpretability of such animal traces, see Fudge, "Left-Handed Blow," 5. On the survival of traces of the dead through recording technology, see Friedrich A. Kittler, *Gramophone, Film, Typewriter* (Stanford, Calif.: Stanford University Press, 1999).

35. Abram, *Spell of the Sensuous*, 282n.2.

36. Cary Wolfe, *Animal Rites: American Culture, the Discourse of Species, and Posthumanist Theory* (Chicago: University of Chicago Press, 2003), 90–94.

37. Banting, "Magic Is Afoot," 41.

38. This perspective has elements in common with Alfred North Whitehead's process philosophy; Isabelle Stengers, *Penser avec Whitehead: Une Libre et Sauvage Création de Concepts* (Paris: Seuil, 2002); Alfred North Whitehead, *Science and the Modern World* (Cambridge: Cambridge University Press, 1930).

39. Geoffrey C. Bowker, *Memory Practices in the Sciences* (Cambridge, Mass.: MIT Press, 2005), 19.

40. Edmund Russell, "Evolutionary History: Prospectus for a New Field," *Environmental History* 8, no. 2 (April 2003): 204–228.

41. Fudge, "Left-Handed Blow," 5.

42. Marc Bloch, *The Historian's Craft* (New York: Knopf, 1953), 55.

43. Bloch, *Historian's Craft*, 55.

44. Carlo Ginzburg, "Clues: Roots of an Evidential Paradigm," in *Clues, Myths, and the Historical Method* (Baltimore: Johns Hopkins University Press, 1989), 96–125.

45. Ginzburg, "Clues," 101.

46. Ginzburg, "Clues," 103. For related arguments about the centrality of hunting and tracking as the origin of narrative and as an epistemological model for other areas of human activity, see Liebenberg, *Art of Tracking*; Chamberlin, "Hunting, Tracking and Reading."

47. On the subject of disciplinary distinctions, Bloch writes: "What is it that seems to dictate the intervention of history? It is the appearance of the human element." Bloch, *Historian's Craft*, 21.

48. Ginzburg, "Clues," 119–123.

49. Ginzburg, "Clues," 101.

50. Derrida, *Animal*, 112.

51. Ernest Thompson Seton, "The Oldest of All Writing—Tracks," *Country Life in America* (December 1909), 169.

52. The early history of bird banding and observational networks of amateur and professional birdwatchers in the United States is described in Mark V. Barrow, *A Passion for Birds: American Ornithology after Audubon* (Princeton, N.J.: Princeton University Press, 1998), 154–180.

53. On the "politic semiotics of articulation," see Haraway, "Promises of Monsters," 83; on technological systems (or the "ecology") of traces, see Bowker, *Memory Practices*, 4–6.

54. On ecology as a mode of control of both animals and humans, see Peder Anker, *Imperial Ecology: Environmental Order in the British Empire, 1895–1945* (Cambridge, Mass.: Harvard University Press, 2001).

55. On archives as collections of traces, see Bowker, *Memory Practices*, 4–6.

Mobility and the Making of Animal Meaning
The Kinetics of "Vermin" and "Wildlife" in Southern Africa

CLAPPERTON CHAKANETSA MAVHUNGA

MICROBES, INSECTS, BIRDS, AND WILD ANIMALS (HEREAFTER "THE ANIMAL") HAVE played a key role in influencing humans to make contingent environmental decisions in the Limpopo basin of southern Africa from the mid-eighteenth to the late-twentieth centuries. This exploration is part of a larger project examining human deployment of weapons of war—principally guns and poisons—in ways that level different bodies big and small into one ontology of pesthood. The status of being a pest (an organism whose characteristics humans regard as injurious or unwanted) is the ultimate designation of living things, including humans, as animals. Because some humans assign other humans the status of being animals, it needs to be noted that being human is, in itself, a subjective status either self-appointed or granted by others favorably inclined toward their object of designation, while alienating others.[1]

Subjective designations of pesthood that clear for destruction certain bodies—of microbes, insects, fauna, and humans—are usually sanctioned by those with political and technological power. They exemplify, and move beyond, Foucault's humancentric notions of biopower.[2] While meaning it as a technology of power, Foucault limited such power to human power over human bodies, but we need not end there. Beyond the obvious extension of human power over nonhuman living and nonliving bodies, it is possible to see the human as a subject of nonhuman power, to expand biopower into nature's power over us.

In this chapter, in which what is at stake is life itself, it remains true that technology and politics mutually lend power to each other, thereby shaping the relations of humans (as species) toward other species. This work is violently opposed to confining the configuration of power within the semiotic reference or human discourse, that is, merely seeing the othering of the other as language of assigning labels to things that cannot speak back (loudly enough). I leave it to others to bear witness to the good things about the animal that leads them to humanize it into a "she" or "he"; my focus is on moments when "pest" gains or loses the "s" and on the role mobility plays in this process of name-calling and the cuddly or violent actions that it portends. The chapter does not see labeling as a sign of hegemony or being labeled as a sign of weakness. It resists the impulse of scholars to start their exposition at this secondary stage of what is obviously a chain of actions, with labeling getting on an already moving bus and risking being thrown off or left behind altogether.[3]

As the linguistic essence of the animal, the othering of the other is a sign of an entity that materially and visually avails itself for designation, not like a meek lamb led to the slaughter (even a sheep being sacrificed possesses agency when its properties trigger human cognitive and spiritual beliefs in its material plausibility to cleanse), but being too pestilent to the labeler's existence to be ignored. We cannot simply end with how the cultural environment codifies how we imagine the animal, as if animals are always at our imaginative mercies.[4]

The animal, I propose, is first author in its designation as an other and is not coming to discourse as an innocent child. It has a materiality that lends itself to the human readings that we call subjectivity. Then again, so do nonliving things like precious stones, sparkling water, tough metals, and stuffy air. The difference is that the animal possesses a consciousness derived of brain and nerve; hence its kinetics, its mobilities, are to be talked about as purposeful material traction of body through space and place. These purposive kinetics are open to the subjectivities of the human as a species: either they portend good things and must be encouraged or ill tidings and must be destroyed.

This, then, is the late-nineteenth- and late-twentieth-century history of the lands either side of and inclusive of an international southern African river called the Limpopo, especially from its confluence with the Shashe to just before it pours into the Indian Ocean. Since 1998, conservationists have been urging (and sometimes even dragooning) the governments of South Africa, Mozambique, Zimbabwe, and Botswana to avail those parts of their countries riparian to the Limpopo for the creation of one vast transboundary conservancy where wildlife can simply roam without encumbrance (see figure 1).

This pressure to conserve nature is the second focus of this chapter: wildlife as "the good animal"—"wild" because it lives in forests whose trees were never sown by humans, which grow naturally because humans have fenced the boundaries to prevent "animal humans" (humans who behave like bad animals) from hunting the animals for meat or cutting the trees for energy and construction purposes. Wildlife is game—that is, animals whose lives are spared for their amusement or pastime value to specific humans who have enough money to come and be amused by their existence. The good animal, therefore, lives in a game reserve, a place reserved for animals kept in confinement (inside electrified and/or barbed wire fences) so that they stay within sight of the client who pays to derive amusement from them. Otherwise the game reserve is also called the game park, a place for the recreation and enjoyment of specific humans for whom the very idea of seeing an animal is recreative and joyful.

How exactly did we end up here, at a stage where a version of the animal as wildlife to be conserved at any cost becomes more dominant over the animal as vermin to be annihilated at first sight? First, we need to appreciate a period in which wildlife did not exist, when the animal was vermin. After that, we can then look at a time when vermin no longer exists, when the animal is wildlife. These transformations are footprints of those in political and financial power who determine the species they can live with and those that must die for humans to live. By now it should be clear that what is often called Nature with a capital N is no more than a footprint of ground stepped upon by a specific mood of history. Nature is merely what political cultures set aside as worth sparing or exterminating. Both "the good animal" (wildlife) and "the (bad) animal" (vermin) are markers of Nature writ large.

This chapter is an attempt to theoretically make sense of heavily empirical work on the veterinary reasons why the transfrontier scheme along the Limpopo can now be possible at

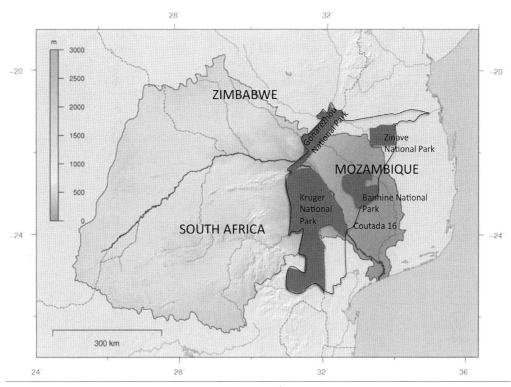

Figure 1. The Great Limpopo Transfrontier Park in southern Africa. (Source: Clapperton Mavhunga 2010, adapted from wikicommons)

the beginning of the twenty-first century, but failed at the start of the twentieth. It serves the purpose of advancing theoretical inquiry into the calculation of my own species' self-portrait as human while designating species that do not look like us as the other. The chapter brings into conversation—and rejects—the three temporal moments often used to define lands subject to colonization by Europeans, namely the precolonial, colonial, and postcolonial. These labels make little sense because on the ground the abrupt dates used to demarcate time elide the temporal fluidities that one can see clearly by training the analytical lens on the mobilities of bodies and the play of mobility itself as agent.[5]

UNDISCIPLINING THE ANIMAL:
ESTABLISHING EPISTEMOLOGICAL CONVERGENCES

Repudiation of categories trained on the mobilities and sedentarization of political actors and actions can only succeed by allowing them to float within a constellation of other actors (actants) (specifically technology, animals, and plants), not by silencing political milestones (changes in government). To do so effectively, the chapter combines perspectives in the fields of science, technology, and society (STS); animal studies; environmental history; and African studies to rethink human-animal interactions. The major epistemological challenge is how one can bring together four disciplines that emerged out of context-specific epistemological imperatives into an effective theoretical framework for understanding the meaning of the animal.

Science, Technology, and Society (STS)

STS stemmed from a Cold War determination to escape technological determinism—the idea that technology drives society, specifically the interaction between humans and nature. To the contrary, science and technology were socially constructed.[6] For some STS scholars, it is not enough to simply state the social construction of technology and science, as if technology and science do not also shape society. What is more fundamental, these scholars argue, is co-construction or coproduction—the dynamic interaction between science and technology (technoscience), technology and society (sociotechnical), technology and politics (technopolitics), technology and culture (technoculture), and technology and nature.[7] These works enable us to explain the work technology does in human interactions with the animal.

Significant strides have been made in STS on the subject of what the animal itself does that influences human actions. The groundbreaking work was without doubt Michel Callon's article on scallops, which created room for nonhuman living things to be included as actants in the assembling of the social.[8] There is Donna Haraway's work on the cyborg—a cybernatic organism, a hybrid of machine and organism, something that is created between fiction and lived social reality, and is always situated.[9] Alongside Haraway, Harriet Ritvo (who straddles STS and animal studies), especially in her *The Animal Estate*, has examined the animal as a reflection of social and cultural meaning as well as a site where human contests over status, social discipline, and empire are fought and resolved. Among these contests are veterinary campaigns, the creation of zoos and menageries, and big game hunting. Ritvo is not willing to concede that the animal comes to the table merely as a construction of human imagination and technological work. To the contrary, the animal produces and reinforces boundaries between social classes and racial groups. Ultimately, colonial conquest—or colonizing—was about the European imperialist coming out on top over nature.[10] Yet one feels that existing work blurs the boundaries between domestic and wild animals too finely.

Animal Studies

No other body of work comes closer to humanizing the existence of the animal than the field of animal studies, which grew out of an activist genre seeking the recognition of the rights of the animal. This was a significant step: it meant that living organisms that could not open their mouths to articulate—save for hissing, mooing, braying, and wincing—their feelings and political views could actually be given rights by certain members of the human species not too dissimilar from human rights. This radical step emerged out of ethical questions surrounding killing, eating, and experimenting with animals. It was a quest to understand animals as beings-in-themselves that do not owe their existence to our knowledge or recognition of them.[11] The philosophical manifestations on the question of the animal had for a long time focused on our moral obligations to nonhuman beings. Indeed, Martin Heidegger implored all of us to take responsibility in respecting of all forms of life,[12] following on Emmanuel Levinas's work on the ethics toward the nonhuman.[13]

Yet since the work of Jacques Derrida, even the human versus animal dichotomies that Heidegger had deployed suddenly look suspect.[14] The hard sciences have elaborated on Darwin's gradualist continuism, while in the humanities and social sciences markers of "the human" have been found present in animals or expressing themselves in different, even animalistic, forms. As Matthew Calarco has already noted, just "what is supposed to constitute humanity, when the notion of humanity is undercut," is as much in doubt as the concept of animality itself. There emerges a conundrum: to redraw the contour line, or to abandon the concept altogether. Calarco is convinced that under such circumstances "*the human-animal distinction can no longer and ought no longer to be maintained*" [his emphasis].[15]

So what is to replace the animal? Calarco, following Derrida, calls for a reframing of "the question of the animal"[16] so that it does not endorse most philosophers' limitation of the question to the content of animality or "the being of 'The Animal.'" This shift beyond the essentialist framing of the animal calls into question Levinas's own concerns with ethics. As Calarco paraphrased it, "the face of the Other cannot be delimited *a priori* to the realm of the human . . . an animal whom I face and by whom I am faced and who calls my mode of existence into question."[17] This elusive dichotomy is what forced Giorgio Agamben to declare the human-animal dichotomy in Western theory as, to paraphrase Calarco, "the chief obstacle for a postmetaphysical concept of relation and community."[18]

In considering Calarco's insights, one still finds a view of the animal that does not consider it as vermin whose pestilence—its trouble-causing mobilities—to humans constitutes (if its kinetics are traced as a subject of inquiry) the best way to observe the limits of boundaries and boundlessness. It therefore seems to me surprising that *The Animals Reader* published in 2007 did not deal with animality as pesthood—a kinetic of pestering. The volume contains thirty-five chapters spanning topics on animals as philosophical and ethical subjects; reflexive thinkers; domesticates, pets, and food; spectacle and sport; symbols; and scientific objects. None of the stellar cast found the subject of the bad animal—the beast—as worth considering, specifically the way animals themselves actually make meaning of themselves when humans interact with them, or how they make humans humane or animalistic through the kinetics of interaction.[19] Of all the chapters, the most relevant for this chapter is Jeffrey Moussaieff Masson and Susan McCarthy's "Grief, Sadness, and the Bones of Elephants," which allows this present work to go further beyond animals just feeling grief, loneliness, and imprisonment, to say that insects, birds, and animals—unlike trees—move purposefully and encounter the human as such.[20]

Close attention to how the pest intrudes our human space is important to the crafting of the response befitting of a pest, to squash or annihilate it. At the same time, by a law of symmetry, it helps to watch closely the mobilities of humans into the space of the animal, especially if such movement is uninvited and unwanted, wherein we become the pest. There is a third angle: our experience—itself a function and product of mobilities—of the destructive movements of animals in *our* space enables us to tell when our fellow human beings descend into mobilities resembling the animal. Here I refer not to the cuddly animal we see as a pet but the beast, the serpent, the bird and animal of prey whose presence bodes ill for our property or us, against which weapons designed for pests—any kinds of pests—must be deployed. The process of "becoming animal," where "all forms come undone,"[21] is a dialogue between two opposites—what we see as the best in us (our humanity) and the worst in other living species (their animality).

Before we throw away the distinction between animal and human, it is crucial to explore whether the zone of distinction is now redundant, replaced instead by a zone of indistinction.

I doubt it. It never will; if it dies, so does humanity. The best place to view this, even under the changed circumstances where vermin is now wildlife, is the game reserve.

African Environmental History

STS and animal studies are the analytical conveyor belt I use to take the question of the animal into African studies—specifically environmental history—as an actor category. African studies emerged out of a concern to give Africa a history, especially since the nineteenth-century German philosopher-historian Hegel had declared its people to have none. As it turned out, right from its foundations African studies—history and anthropology in particular—was almost exclusively a narrative of what people were doing to/with technology, animals, and their environment. So long as African agency could be demonstrated, the role of nonhuman things was not a priority. The terrain is slowly changing, largely with the interest of STS scholars in Africa.[22]

African environmental history is a more recent addition to a growing field of African studies. It drew its inspiration from the environmental history emerging from the rise of American environmentalism in the 1960s–1970s. Like STS today, its empirical spread to other parts of the world grew out of the need to contextualize and universalize the theories, temporalities, and spatialities that had started emerging from Western-centric examinations.[23] No wonder the major influences have come from these Western scholars. William Cronon's *Changes in the Land* (1983) first shattered the myth that precolonial landscapes were pristine wildernesses never touched by humans or culture until Europeans and their technologies arrived.[24]

In 1986, Alfred Crosby published *Ecological Imperialism*, arguing that Europeans were successful in colonizing other humans because they transplanted their crops and animals at the expense of those on which indigenous populations depended. In calling these spaces of conquest "Neo-Europes," Crosby was talking about North America, South America, Australia, and New Zealand; Africa and Asia were not included in the discussion. Efforts of scholars like William Beinart in southern Africa;[25] John McNeill, whose work connected Africa, southern Europe, and the Middle East via the Mediterranean;[26] and James McCann opened up space for the introduction of environment to studies of African history. Whereas scholars like McNeill and Beinart adopt more regional approaches, McCann has ambitiously opted for a more continental thrust.[27]

Generally speaking, these environmental studies of Africa consider what people are doing to it, with it, and within it. Nature is not an actor, perhaps because in most of these studies there seems to be a tacit yet palpable reluctance to grant agency to nature lest it casts Africans as its hapless captives rescued by the timely intervention of Western technology. The dilemma is how to locate the agency of nature without subtracting from that of the (human/colonial) subaltern. By the time Africans transcend from being the political subaltern to nationalists who acquire political independence, the interest in nature's agency has already been lost. Environmental histories of Africa become like much of African history: a narcissistic human gaze on the train of history.

The Limpopo basin is no different. The most recent work on the Zimbabwean side—William Wolmer's *From Wilderness Vision to Farm Invasions*—examines two key human activities, conservation and development, in the conceptualization of this landscape as wilderness. Wolmer is interested in what humans are doing, first to each other as a function of power relations between the state and local inhabitants, and second to the land (inclusive of its plants, animals, and soils) as a function of power relations between two species, the human and the (wild) animal.[28] What becomes of the landscape, and who comes to own or benefit from the land, depends on who is in power over animals, locals, and the soil. It is a pity that the voices of the land, the animals, and the plants are muted into silence in order to enable that of Africans to be heard: indeed, that is why nature is but just that—a wilderness vision. It is the same story south of the Limpopo River.[29] The Limpopo River, the national border between South Africa, Mozambique, and Zimbabwe, is epistemologically worshipped like a god(dess); scholarship is fearful of crossing it, even as nature does.

When animal and nation are in conflict and the mobility of the former is impossible to enforce, or when the human subjects of the state move about in transgressive ways befitting of animals, the state resorts to quarantine or annihilation. The animal renders itself deadly to the nation because it (unintentionally) carries microbial pathogens or habits destructive to the nation's political, economic, health, or social interests. Human subjects of the state become pathogenic to the state by behaving animally. The abandonment of earlier transboundary conservation plans in the 1920s–1930s stemmed from human-animal movements across the Mozambique-Rhodesia (now Zimbabwe) border, that carried the tsetse fly species *Glossina morsitans* and *Glossina pallidipes*, which themselves carried trypanosomes, which caused sleeping sickness in cattle, the mainstay of Rhodesia's agro-based economy. The Rhodesians could not go through with plans to declare Gonarezhou a game reserve. The South African government could not countenance a tsetse-infested game sanctuary on its borders, with animals and people traversing back and forth, carrying tsetse fly and trypanosome into Kruger National Park and its village neighborhoods.[30]

In a part of the world where nature, or the animal and its space (the park), was created through the colonial state's displacement of people inhabiting the land, it became inevitable for environmental justice questions to form a basis for black political resistance: all that nationalist movements needed to do was to point at where blacks now lived, and ask them where they used to live, and most of the time villagers would support them. If not, the guerrillas resorted to violence to make villagers see the dehumanization of colonial land dispossession.[31] This is also why environmental histories of Africa cannot escape the cleavages defining the history of Africa for decades as sociopolitical. Colonial states used environmental science to justify policies that disadvantaged black people, such as livestock culling, land dispossessions and resettlements, and prohibition of colonial subjects from hunting.[32]

Inevitably, the locus of inquiry in the wake of political independence has been whether blacks have regained the advantage they lost through colonialism. This research agenda straddles both the quest to know and to use knowledge known to transform the situation through development and conservation. It is hardly a coincidence that most scholarship on the Zimbabwean side has come from development studies and is interested in the historical loss and (potential for) the creation of sustainable livelihoods. In these narratives, the animal is at the mercy of human design. Whatever the animal does for itself, its behaviors are calibrated as risky actions eliciting human responses; the credit for the outcomes goes to humans.

The animal's mobilities manifest themselves as veterinary disease—a threat to the livestock-based economies of communal landowners and beef exporters to the European Union (EU) on either side of the Limpopo River border.[33] The discussion on the Zimbabwean side of the Limpopo has shifted to the land reforms under way there, especially the internal displacement of animal and human inhabitants as Zanu (PF) officials and supporters of Robert Mugabe, the country's only ruler since political independence in 1980, invade and occupy parts of Gonarezhou and entire private game reserves. The human-animal boundaries we have taken for granted for so long have been thoroughly transgressed: people and animals become coresidents of both game reserve and rural countryside.

On the South African side, the human-animal connection is deeply informed by the success of the Makuleke people in reclaiming from government the land from which they were forcibly removed in 1969. This has since triggered a wave of land claims and restitution cases. A quid pro quo had to be made whereby local communities would only get their land inside the game reserve if they agreed to lease it back to the National Parks Board so that Kruger would retain its boundaries and function as a national park while recognizing the injustices of history.[34]

Beyond the riverine ecology of the Limpopo, two more works form the immediate backdrop to the second section of this chapter, insofar as they connect nature and identity, in the first instance, and identity and power, in the second. Terence Ranger's view of Rhodesia's Matopos National Park in the twentieth century goes beyond nature—the animal—as natural resource available for human livelihood, to nature as a heritage demanding of certain ethics toward it—ethics that, as Cronon had shown in the case of New England, predated European colonialism. The Department of National Parks saw Matopos as "a uniquely fragile ecosystem" needing protection from human-animal destruction; Matopos's African inhabitants saw a centuries-old human-modified landscape sanctioned by history, religion, and custom. Ranger found it impossible to separate African identity from the proximity and symbolic power of Matopos as a spiritual shrine.[35] In the case of the middle and lower Limpopo, Isak Niehaus has found that "the content of ethnic groups has no *a priori* existence, stability or coherence."[36] Indeed, Niehaus showed just how often such identities "acquire salience under conditions of inequality" and frequently defy the classificatory grids of the state, such that, even under colonial subalternity, people still defined themselves, the state, and nature in their own ways.[37]

The weapons with which the state tried to tame the wild animal—of the human and faunal variety alike—was the gun. Niehaus and Ranger framed identity within a human domain of identity formation without breaking into the power of nonhuman actors like guns and animals to identify. That is what animals do to poachers, who use certain tools (for example, guns) to achieve ends (for example, kill animals). As Bruno Latour reminds us, the gun is "a tool, a medium, a neutral carrier of human will. If the gunman is a good guy, the gun will be used wisely and will kill only when appropriate. If the gunman is a crook or a lunatic, then with no change in the gun itself, a killing that would in any case occur will be (simply) carried out more efficiently."[38]

Humans who use guns for bad purposes (according to the state) are akin to "human animals"—people who behave in animal ways, who shall be treated as such. Technology is what humans behave with. Nature—the animal and its environment—is the subject, evidence, and venue of the human's animal behavior. It is a fair critique to say that scholars are studying environmental histories of Africa—adopting themes borrowed from Crosby, Cronon, and others to African history—but not necessarily the African environment itself. If they were, then the nonhuman would not necessarily have to first come into contact with or beg the human to deserve attention as a historical subject. It is as if anything that does not affect humans is not historical.

VERMIN: THE BAD ANIMAL

Humans summon technology as an antidote against pathogens that threaten them. Whether it is the economy and they seek more wealth, arduous work that they want to simplify, difficult life that could be better, or illness that could be healed, technology is summoned to cure. In this section, technology—the gun—is invoked to deal with veterinary disease; tsetse fly mediates the relationship between the animal and the human (and between the state, colonial settlers, and local inhabitants) and defines a role for guns as antidotes in the absence of any medicinal cure.

Guns and Deadly Microbia

The state of historical knowledge on veterinary technologies—and veterinary knowledge in general—in southern Africa is only just beginning to gather momentum, focusing largely on the colonial period.[39] This miserly record is not for lack of sources. Almost every precolonial European traveler mentions the tsetse fly belt—inclusive of Gonarezhou—while the fight against the pest itself generated some of the most amazingly detailed archives on twentieth-century Zimbabwe. Probing this terrain enables us to complicate how, by default, the tsetse fly protected lands inhabited by locals that Europeans could easily have seized had these lands been clean and not infested. The tsetse fly also kept away many European hunters on horseback, thereby preserving the animal—principally elephants—from massive slaughters for their trophy.

Once the lands were cleared of the tsetse fly, the animal became subject to two human designs. One was the impulse to satisfy white colonists' senses of class, racial superiority, and masculinity, and to meet nutritional, masculine, or spiritual imperatives of local inhabitants; or in resistance to forced removal from their animal-rich lands. The second impulse was to remove all previous residents from their animal-rich lands and declare such lands as game reserves for purposes of wildlife conservation, using models of nature borrowed from Yellowstone National Park, Niagara Falls, and Kruger, among others.

For much of the period, the key strategies against the pathogen and the animal that carried it were game slaughter, cordons, relocations, and laboratory innovation, mostly trial and error in uncharted waters.[40] This Western regime of veterinary knowledge did not openly acknowledge the role of Africans in making expertise, despite extensively using African medico-environmental knowledge that had inducted nineteenth-century expeditioners on the typology and spatiality of tropical disease, thereby making a bad situation worse.[41] A good number of sources deal comprehensively with the disruptive effect of human and livestock movement, but not with the instruments of disruption and dispersal of animals and people who became not simply colonizers, soldiers, victims, and refugees but also carriers of pathogen.[42]

My immediate concern is with respect to the instruments used to protect society—including its assets, collective or individual—against microbes. Specifically, I want to examine the use of guns as medical tools against pathogens, parasites, and hosts where vaccines had no assured outcomes. The focus of African historians and anthropologists of medicine on bacteriological aspects overlooks the role of soldiers, policemen, and game scouts in veterinary disease control, as if they were merely ancillary to laboratory science. Yet until vaccines or prophylactics were developed, guns, poisons, and foot patrols were used to vaccinate the landscape owing to their portability, lethality, deception, and mobility. Even when synthetic pharmaceutical drugs were invented, these methods still complemented laboratory science.[43]

The gun and tsetse fly are vectors of death, even if they do not kill; it is the bullet (propelled into flight by a human pull on the trigger) and the protozoan (released from the body of the tsetse fly via its proboscis biting into animal flesh) that causes death. Both the gun and the fly are portable; they acquire agency through mobilities of vectors like the animal and the human. Both are harmless without their ammunition, the bullet and trypanosome, respectively. Yet while the gun's use is determined by the bearer, the tsetse fly has a life of its own. Portability, therefore, makes the two organisms malleable to the whims of their carriers, man and animal, respectively.

Because the tsetse fly and guns defy containment and transgress human-nonhuman borders, they are network builders between the forest (the domain of the animal) and the village (the domain of the human). From the state's point of view, both domains are assets when contained and dangerous when they are not. Villagers going into the park to hunt without permits are poaching: they must be arrested. Animals leaving the park to forage into the village for any reason become problem animals or vermin. They must be destroyed lest they spread veterinary diseases.

In the view of aggrieved villagers, the national park fence is a sadistic monument to their forced removal from their lands to create parks, which the new government has maintained. The original fences *kept out* veterinary diseases from domestic spaces, and when the game reserve was established, these fences were reinforced to *keep in* or protect animals through statutes and guns. Whether it is the animal or the human, there seems to be too much emphasis on technologies of separating the species from each other, yet under European colonial rule both occupy the same ontological plane as the problem animal when one strays into the domain of the other. One is the poacher in the park; the other is vermin in the village.[44]

Mobility, Colonization, and Microbia

Mobility defines whether the animal or the human becomes an animal in the sense of behaving out of place. At a historical level, mobility enables the production of a map—which is why Jessica Dubow called it "a place found by way of the body."[45] Hence precolonial European travelers in the process of hunting, preaching, trading, and exploring compiled accounts and sketches that would carve open ambitions for colonizing Africa. In the last twenty years of the nineteenth century, European monarchs, governments, and corporations undertook military expeditions to conquer these explored lands, acting on the basis of treaties these travelers had signed. After tricking or conquering African rulers into subjecthood, these Europeans set about transforming lands their countrymen had previously just fleeted through into permanent sites of residence. Where guns had relied so much on human movement to conquer the colony, now the nascent state deployed them against rinderpest, the tsetse fly, and later insurgency. And where some movement conveyed scientists and pathogens to new areas, other movement prevented them from doing their work.[46]

Movement, therefore, transformed the surroundings by breeching old and making new borders favoring the new settlers. The work of Helge Kjekshus in Kenya has illustrated clearly how colonization was a process of upsetting precolonial, antidisease stratagems and inserting new ones deemed to be scientific.[47] Colonial border creation split ecological habitats established for decades, centuries, or millennia into different states or administrative districts. International borders cut straight through people's villages and animal haunts, deliberately pathologizing the normal, kinship-based mobilities. By their own movements, driving infected stock along, Europeans carried the rinderpest pathogen from the steppes of the Balkans into northeast Africa as live cattle rations for their colonizing forces (1888), and through their wagon routes south as far as the borders of Cape Colony (1897).[48]

We know from historical documents that precolonial inhabitants of the trans-Limpopo basin had their own veterinary measures separating the animal from the human spheres. To cite just one example, entomologists C. F. M. Swynnerton and John Ford mention that the Gaza rulers of nineteenth-century Gonarezhou manipulated migration as a mechanism to escape harsh seasons and deadly insects, principally the anopheles mosquito (which causes malaria) and the tsetse fly (deadly to cattle). When introducing cattle into the northern fringes of Gonarezhou captured farther south, the Gaza ruler Mzila (in power between 1861 and 1884) ordered his people to *sondela enkosini* (draw near to the king). Through this population-control measure, Mzila cleared vast tracts of buffer zones by overgrazing and deforestation. He then left certain areas in his domains unsettled as game reserves; inside them, hunting was banned, lest it scuttle and scatter the animals and their pathogens into the clean spaces.[49]

When Mzila's son and successor, Ngungunyane, retired with all his people to the southern coastal area near Maputo in 1889 to escape British and Portuguese advances into his domains, these veterinary boundaries collapsed. The tsetse fly started moving in tandem with the dissipation of ecological boundaries: as the animals roamed into erstwhile villages, the deadly insect found convenient transport on the bodies of what was, after all, also its primary food source. The partition of the area into British and Portuguese territories between 1891 and 1895 compounded the problem even further. It simply split these prior human-modified animal habitats and human settlements and saw subsequent movement across them as transgression.[50]

Colonization as Microbial Mobility

In 1896 rinderpest arrived. The strain, which killed cloven-hooved animals, is a typical case of ecological imperialism, although it was a double-edged sword affecting both the settling and preceding inhabitants. Rinderpest is an interesting site for studying the mobilities and meaning of microbes in the body system of moving animals, infecting other mobile animal life around and beyond it. It transforms completely our understanding of the animal. The domesticated animal (livestock) suddenly becomes vermin, a source of pathogens that is no different from, if not even more dangerous than, the wild animal. It must be put to death to prevent contagion; in fact, it is—by virtue of carrying a bug inside its body—contagion. Europeans interpreted rinderpest as a microbe, but did not have a pharmaceutical cure. The antidote against the bug was the gun and the patrol cordon: to eliminate the pathogen, the fumigation teams slaughtered its carrier, cattle and other livestock.[51]

Africans interpreted it as God's retribution against whites. For them rinderpest was not the disease. Rather the curse was the white man who must be chased out to cleanse the land.[52] The cure was to destroy the presence of whites and restore the normal order of things. To fight this disease, local rulers and their people, especially the Ndebele and Shona, took up arms.[53] Indeed, colonial officials at the time correctly pointed out that in African cultures, there was no such thing as a natural occurrence in the Western scientific sense. There was a cause behind the cause of this pestilence. So what was causing the cause?[54]

In these two very different interpretations of what the disease was, we also see two interpretations of the animal. For white settlers, rinderpest was a novel clinical affair for which they had no antidote other than the deployment of guns to enforce barriers between infected and uninfected cattle herds. Mass slaughter and burial of livestock ensued, including but not limited to those belonging to the colonized. In so doing the settlers only added more proof of being the disease. The colonized interpreted this stamping out (with some truth) as a well-calculated economic disempowerment to force them into wage labor on white farms and mines. In the antirinderpest operation they saw clearly the mark of the beast. The white settlers' "real object [was] to remove them from the face of the earth by killing their cattle and leaving them to die of starvation."[55]

Ways of interpreting the animal designated what sort of instruments were necessary to tame it. For both, the key was to use mobility to freeze the mobility of the pathogen— permanently. The new settlers deployed guns to freeze cattle and human movement, hoping to prevent infections in new areas. The colonized human subjects used rebellion as a lethal kinetics to annihilate their own version of the virus: the white settlers contaminating their livestock with their diseases, causing hunger and poverty through killing their cattle and suppressing their right to be left alone.[56]

In areas where the tsetse fly and wild animals were dominant, the more relevant discussion is how the bug wiped out the animal. Left without its foremost food source, the tsetse was eliminated in many parts close to the wagon route from British Central Africa to Cape Colony. This is where Crosby's "ecological imperialism" argument is correct: rinderpest's elimination of the tsetse fly intensified human beings' domination of the land through the application of technology. Some places morphed into towns, others into livestock ranches (most cattle being imported) and crop farms, yet others into reserves where the animal was relocated and preserved.

The Tribesman

"The reserve" was a place to preserve two types of animals. The "game reserve," now de-flied and cleared of the colonized humans, became a place exclusively for untamed animals, with a fence on its edges to preempt their rebellious movements into the countryside where they might become vermin. The native reserve, cleared of all whites save for priests, administrators, and outcasts, was established as a place for the "tribesmen," wherein they were quarantined and monitored (by Native commissioners and their staff) to prevent them from contaminating white space through rebellion and backwardness. Carefully measured dosages of civilization could then be fed to them, making them just enlightened enough to take orders while not endangering white civilization. Three spheres of spatial differentiation emerged: the domain of (white) civilization (farms, mines, towns), that of "the tribesman," and that of wildlife. For some officials, the domain of "the tribesmen" was that of "wild Africans," which, like wildlife, would be very attractive to European tourists. That was why the district commissioner for Nuanetsi—covering the northern bank of the Limpopo in the 1960s—agitated for the local Shangane inhabitants to be allowed to stay inside the emerging Gonarezhou Game Reserve so that tourists could come and "study wild animals and 'wild' Africans."[57]

The ability of the state to engineer "the tribesmen" toward the door of civilization depended on how physically distant the state was from the subject of its discipline. State authority could only be felt if government officials were visibly and materially present. Yet on the borderlands where malaria, trypanosomiasis, and a complete lack of transport infrastructure were the order of the day, the state could only patrol once a year in winter. Illicit labor recruiters for South African mines and poachers filled in the void and literally became the state between 1890 and 1930, using the inducements of meat and money as violence to obtain local villagers' connivance.[58]

Poachers and Human Traffickers

Just why and how these men's poaching and recruiting peaked during the tsetse-free period illustrates how freedom from the tsetse fly created conditions for certain technologies to thrive in ways they would not before rinderpest. This use of guns by individuals who did not answer to any of the three neighboring colonial states—Southern Rhodesia to the north, the Union of South Africa to the south, and Portuguese East Africa to the east—was in itself a kind of disease that had to be controlled. Yet like rinderpest and tsetse fly, it occurred when no antidote was in sight: new colonies had no clear modus operandi for transborder security cooperation. The earlier response to vermin was applied: state guns were deployed against the spreaders of pathogens.[59]

In a time when veterinary pathogens posed limited threat, the state entertained the idea of elevating the animal from vermin to wildlife, and the coexistence of livestock and game. This brief honeymoon took place in the trans-Limpopo in the 1920s to mid-1930s out of alarm that unregulated hunting was virtually exterminating all wild animals on either side of the Limpopo. While foot-and-mouth disease (FMD) and East Coast Fever (ECF) were both causing havoc on the large ranches further west such as Nuanetsi and Liebigs, their levels of virulence could be controlled through cordons. The implosion of the tsetse fly, riding on the backs of animals returning to the haunts that rinderpest had decimated in 1896–1897, jeopardized plans to declare a transboundary game reserve linking contiguous territories in Rhodesia, Mozambique, and South Africa. The empirical aspects of that development have been discussed elsewhere.[60] In the following section I make a few theoretical remarks on the outstanding empirical fault lines.

Wildlife

Game reserves are moments; they are occasions, moods at exact points in time, and like all moods and occasions they are not necessarily permanent. They are the moods of the human mind, the human that is in power, whose ideologies toward nature determine its destiny. The imperative to destroy the animal, to let the animal be destroyed, or to protect the animal illustrates the changing moods of the colonial. Destruction could even be read as a concomitant step toward the purification of nature, to winnow out bad nature and leave only the good, the distillate.[61] This cost-benefit strategy forced the three neighboring governments to take steps toward transboundary game management, more out of fear of pathogens spreading than for the love of the animal. Those who proposed wildlife tourism as a land-use model

(1930s–1950s) to justify game conservation failed precisely because game and cattle were incompatible. FMD outbreaks in 1934–36 and the advance of the tsetse fly from central Mozambique from 1938 onward solidified these fault lines.[62]

The use of guns against helpless animals just to quench an appetite for blood and carcasses that men then worshipped as sport and trophies struck some Europeans at home and in the colonies as barbaric and bloodthirsty. The morality of science lay in the way a cure would sustain and multiply life. Hence stopping the antitsetse operations (that is, game slaughters) by fencing and vaccination represented a way to reassert the moral conscience of whites and confirm the superiority of their civilization. The moral obligation to "the animal" turned out to be more urgent than the moral obligation toward "the other animal"—"the tribesman." The game reserve was being created by displacing "the tribesman." The hoteliers and naturalists in favor of game conservation did not consider any moral—let alone economic—conscience toward the white cattle ranchers who saw the game reserve as nothing less than a vermin reservoir.[63]

The mobilities of insects and invisible pathogens determined whose moralities the government could authorize or sacrifice. The outbreaks of East Coast Fever, FMD, and trypanosomiasis were a function of micromobilities or mobilities inside mobility (inside the bellies of animals constantly on the move) that designated the preservation of the animal as an immoral act.[64] At the same time, control mechanisms directed at preempting or restricting the animal's movements were deemed gallant feats in the protection not just of livestock but also the preservation of the white civilization settlers were designing, for themselves and for "the tribesman." Native commissioners, in charge of shepherding "the tribesman" in the uphill direction of civilization, were more worried about losing land to animals in quantities that would further constrict their mission to "civilize the tribesman."[65] For the burgeoning tourism industry, the imperatives of fashioning the attractions overrode all else; morality lay in ensuring that tourists from overseas, including royal high and mighty, got their money's worth. The fun of the chase was a sine qua non.[66]

The mood change in the 1960s is attributable to a number of important mobilities. The rising white immigration during the federal period (1953–1963) created conditions for diversifying the economy beyond its agro- and livestock base. Tourism acquired powerful new supporters, none more influential than proimmigration prime minister Godfrey Huggins, who saw the benefits of turning human mobilities into wealth, as opposed to thinking that wealth could only come from the ground (that is, from farming and mining). The new turn toward the animal as tourism asset, as opposed to veterinary menace, was celebrated as a triumph of ecological and entomological science.[67]

The battle against the tsetse fly, mobilizing a bouquet of stratagems including game destruction, aerial sprays, quarantines, and laboratory experiments, had been won. The bullet targeted the animal, vector of the insect carrier of the payload (trypanosome). The poison (specifically DDT)—sprayed by ground and airborne teams using small, winged aircraft and helicopters—struck directly at the insect carrying the trypanosome; the death of the host was the death of the microbe. Guns could not be used to exterminate whole forests of animals. That was what quarantines were for. First, fences were strung up on either side of an axis of surveyed land, then the African hunters blazed through with government-issued .303 rifles, shooting at everything that moved within it. In some areas, the bulldozers and African road gangs came in, flattening every giant tree standing, thereby destroying the tsetse's habitat. By 1968, the tsetse had been eradicated, and Gonarezhou Game Reserve was born.

WILDLIFE: THE GOOD ANIMAL

This chapter ends in the present. It focuses on the writer's ethnographic experience on the Limpopo and its south bank. Entering this part of the Kruger National Park is like entering a house—the entrance, the rules, the residents, the sense of being inside. Its northern gateway is Punda Maria Gate. The park can only be entered legally by motor vehicle; for humans, park mobility is motorized. Upon reaching it, the visitor slows down and pulls into the parking space to the left of the road, and goes past the guard into the reception to get a vehicle permit. Once granted, the driver goes back to the vehicle, starts the engine, backs up, makes a slow dash to the guard and shows him the permit. Upon a satisfactory check, the friendly man lifts the booms. The tarmac is wide open to an appointment with *nature*.

The gate is a place where the mobilities of humans are actually regulated. Pedestrians cannot enter because they will be at the mercy of predators; motorists can only enter through the gate. To be legitimately present inside the park, one must be licensed at reception through the issuance of a receipt subject to paying a fee that varies according to one's citizenship. Nationals pay less than citizens of the regional bloc, the Southern African Development Community (SADC), to which South Africa belongs. SADC nationals pay less than nationals of other African countries, who pay less than non-Africans, whose fee to see nature differs according to their nationality. The receipt is to be kept and shown to the guard when one enters and exits the park—to prove that one's appointment with nature is legal. On exiting, the driver is politely but firmly asked to open the vehicle trunk to confirm what the permit shows: that one was in the park for legal reasons and is leaving with nothing that is illegal.

It is hard to imagine a national park or game reserve without its gate. Ironically, the park fence is known not by its water-tightness and ability to keep "the wild animal" inside the fence and the domestic animal outside it. The villagers experience the porosity of the wild versus domestic or the animal-human boundary through the destructive transgressive mobilities of elephants and lions destroying their crops and livestock, respectively. Poachers make nonsense of the gate's role as a human-operated device to keep illegal entrants out: they simply cross under, over, and between the strands of fence far from the gates. The gate can only regulate motorized mobilities but is completely useless against mobilities on foot, be they of animals or humans. Human foot mobilities are left to the game scouts, while the fence regulates animal mobilities—whether it is livestock entering the park or wild animals entering villages.[68]

Mobilities of Animals and Humans

Once one is inside the park, fleeting human-animal encounters begin in earnest. Guides will tell their guest: "To see most game, especially the cats, we need to go pretty early in the morning, at 5.30 am. Or, if you prefer, we can do an evening game drive." This helpful advice usually greets the guest at any private lodge or rest camp, or any backpacker worth his or her salt. What we call a game drive is entirely the human being as hostage to the mobilities of wild animals going about their private—sorry, public!—lives. They may be coming from the waterhole and wandering off to relax in the breezy uplands or just resting by the roadside chewing their cud, or possibly feeding on a carcass, itself a result of a chase in which the strongest survive, the weakest succumb and become meal to others.

It is also because humans are moving—driving—on a road that transgresses the pathways of animals. That is how crossroads are created, crossroads that become the proximity or space of the fleeting encounter with the wild animal. Whenever one vehicle occupant spots an animal, the vehicle stops. Upon seeing this vehicle, those driving behind or from the other direction also stop. Always one is struck by the thought: "The driver must have seen something. Possibly lion, the king of the jungle in all his royal pomposity!"[69] The intuition is always to stop first, train one's eyes onto the faces of the human occupants of the car, to where they are looking, and to follow their gaze until one's own eyes are trained on the prize. Sometimes they may not be watching anything at all, or they are just stopping because they have run out of gas and are calling the emergency police/recovery number for roadside assistance. Always the attraction is fleeting, lasting as long as the mobile animal remains visual. Then it disappears, and the trees, grass, and stones where it had stood moments earlier become just another inauspicious place in the park.[70]

Animal Meaning

The human-animal encounter is a fleeting automotive convergence, transforming a specific space into multiple fragments of international belongings and interpretations. One motorist stops in his Mercedes-Benz SUV, designed in Germany, assembled in South Africa, possibly owned by Avis, the driver possibly coming from Japan, the United States, Britain, or Johannesburg. There I was, a Zimbabwean in South Africa, professor in America, in my Volkswagen Citi Golf, rented from Avis (another fleeting ownership), lining up alongside Toyotas, Renaults, Chevrolets, Mitsubishis, Nissans, Mazdas, Peugeots, Fords, and a gazillion other car models, all gathered here just to see the animal moving around, going through the kinetics of its everyday life.[71]

Here the locomotion of the vehicle is momentarily halted, the engine idling or killed off so that the attraction will not run away or get annoyed. Freed from the concentration on driving, and freezing the potentially distracting mobility of the vehicle, the occupants focus their cameras on the attraction. The driver becomes photographer, shooting the elephant for its giant tusks without killing it—as the old hunters did with guns. Then photographer turns driver and continues the game drive. Months later, the audiovisual images will constitute photo albums and news articles, personal biographies and obituaries. Many more images will even be online.[72]

Engineering the Human-Animal Encounter

The human engineering of road infrastructure has reconfigured the function of mobility in interspecies relations. Central to this transfiguration of nature into a fleeting visual spectacle is the engineering of roads and water systems to be in close proximity of each other. The animals come to drink, and the motorists drive by and see them. From as early as 1898 when Kruger was established, water was deemed critical to sustaining wildlife and attracting tourists. It is no wonder that the park started out in the well-watered parts to the south moving north. The establishment of the park was not a spatial event but a spatiotemporal progression over decades, culminating in the violent removal of the Makuleke people from Pafuri Triangle in 1969 and the extension of Kruger National Park to touch the Limpopo.[73]

Windmill-powered boreholes were central to the project of connecting the animal to its human admirers—the tourists. The movement of the water, itself an outcome of the kinetics of wind, promoted or drew the movement of animals from different parts of the park to come and drink. The construction of human-made water points created new habitats in places that at first sight might be or previously had been very dry. Animals that had hugged the Limpopo for water or trekked long distances to get it could now establish permanent haunts in new places. This created the possibilities for imagining new tourism sites; it became possible to develop lodges cashing in on the attraction that the mobilities of animals created. This led visitors to spend a holiday watching animal mobilities, kinetics configuring workplaces. When we speak of tourism contributing to the economy, we are talking about the convergence of these mobilities. Through them we can map the interactions of animals, wind, terrestrial elements, hydrology, technology, global capital flows, transport systems, and so on. We can even extend Jessica Dubow's definition of a map as "a place found by way of the body" beyond human mapmakers.

The Elephant as Traffic Police Officer

The human road is but one of many in the national park. That of the elephant is distinct: it cannot be missed, and when elephants are crossing, the human road is closed. The elephant's markings are random—the urine, dung, and fresh branches clearly marking out territory. Small vehicles with thin tire thickness have to duck the dung and branches (and a whole tree just felled across the entire road) because elephants eat thorn bushes. These temporary road markings may actually determine what the motorist uses for a road and the specific placement of the wheels in motion.[74]

At which point the question arises: who is transgressing on the road of the other—the elephant or the motorist? The answer, sitting at the crossroads, depends on whose subjectivities we are privileging. The human's road transgresses the elephant's, and the elephant's road transgresses the human's. The answer is often settled by an elephant charge: the motorist either gets out of the way or the motorist is dead (see figure 2). The elephant is akin to a traffic police officer who can halt you at any time, regardless of whether you are on its established path or not. If you do so much as protest, the elephant may prevent you from going where you are going, sometimes even permanently. This raises significant questions about the extent to which human beings are in control if it comes to a point where the forces of nature and the forces of humanity are playing on a level field without any significant technological reinforcements or unfair arbitrations.

Meaning Making

The animal's sovereignty over human mobility is not just physical but spiritual. The ancestors of the local Shona- and Shangane-speaking people on either side of the Limpopo used to say that when you see a certain seldom-seen animal or a bird, like the bateleur eagle (and depending on where you see it, what time, and what exactly it is doing when you see it), something good or tragic is going to happen—the idea being that people just did not see something like that, and it had to represent (something). That is what made them mean something. With the

Figure 2. The elephant as the park's traffic police. (Source: Village Innovations Museum [Makuleke]: Photographic Collections: "Elephant")

coming of overseas tourists, the animal is now being observed just for its visual aesthetics, as a living being.[75]

A prevalent Shona belief holds that when a squirrel runs across the pathway in front of you, it is a sign of danger along the way ahead or at the destination. Some villagers will say it depends on the position of its tail. Among some societies, when the squirrel's tail is up, it is not a good sign.[76] Among others, it is the reverse: hell hath no omen more ominous than a squirrel running with its tail along the ground crossing in front of a traveler. In yet other societies, it makes no difference: a squirrel is a squirrel. Turn back when it runs in front of you.[77] For the Shangane, it is the mongoose. Among some Shona, it is the squirrel. Having squirrels and mongooses run across your path is not a sign of biodiversity; it is a matter of retracing one's footsteps to preserve one's life, or to walk on to one's own demise. It raises a troubling question: what becomes of human senses of hegemony over nature when people cannot move because an animal has moved in a certain way?

Animals Weaponizing Mobility

At a more metaphysical level, the elephant's and squirrel's road-crossing antics illustrate two ways in which animals weaponize mobility. In the elephant's case, movement purveys menace and bodes destruction unless the motorist stops or retreats without any silly negotiations. In the case of the squirrel, movement carries the body as fast across and out of the pathway of the motorized or walking human danger as possible. The flailing or flattened tail acts to increase speed of flight or to conceal the animal's exact location and render it invisible to the potential human assailant.[78]

If one has to dig deep to theorize the weapon function of mobility—to impose the animal's body upon or remove it from endangered spaces—the herd ethic in animals is an example of how animals manage their security through movement. Therefore, one law of animal mobilities is that you are stronger by traveling together and extremely vulnerable when you are alone. The best illustration of this is the way in which prey and herd relate. Buffaloes form all-round defense facing outward. In that position, they can confront any adversary. A You-Tube video taken in Kruger National Park in 2008 tells the story of a whole herd of buffalo charging on five or six lions. The felines were defeated—but not before one was yanked into the air, 2 to 10 meters above the ground. The pride had apparently forced a male calf of one of the buffalo cows into the water. One particularly huge reptile grabbed the calf by the thigh. One lion grabbed the bullock by the neck and began pulling. The calf became the rope in a tug-of-war between amphibian and mammal predator, the one pulling into the water, the other ashore.[79]

Meanwhile, the rest of the mammalian tribe was closing in on a rescue mission. The patriarch of the herd said, "No, I cannot allow these small cats to do this to my son." It started charging, leading the whole herd on a rescue mission, first spreading out to the flanks to close out all escape routes and catching the lions in the bend in the river, leaving only one way out for the feline: the crocodile-infested (sorry, crocodile-rich!) pool. The contest between the crocodile in the pool using its adroitness in water to good effect and the lion tribe on land using the friction and traction of claws dug deep into the ground, rapidly ended with the latter pulling the calf out, rescuing it from the vice-grip jaws of the crocodile at just the moment the herd was closing in.[80]

Then the bovines charged. They rescued the calf and went after the no longer just two but three lions. Two of the felines raced along the water's edge to safety. The main agent provocateur remained and kept pulling as the bull arrived. The lion was yanked into the air and thudded down to the ground. By the time it landed, it beat a rather rapid retreat in the wake of its mates. Herbivore charged after carnivore, hunter became hunted, the normal order of things reversed.[81]

Before we celebrate the warlike proclivities of the animal it is important to see the herd itself as a peaceful movement of animal kin. It is most visible at the waterhole, which exhibits the ethnic diversity of fauna, the equivalent of a human nation if ever there was one. Animals mark out territory, but monopoly is never completely assured—or desirable. Each place in the park is located within a Venn diagram of faunal territory, and any animal's ability to stake its claim depends on mobility. Is the animal fast enough to escape the chase of a lion? Is its herd ethic robust and combative enough to stand and fight?

The predator-prey relation determines which animals are to be found hanging out together. Ordinarily, away from a carcass, each species of predator does not fraternize with anybody else but its own kind. Put some meat in the middle and even kin become enemy. It is perfectly normal to see vultures, hyenas, lions, and jackals all gathered at the scene of the carcass, not in accord but in discordant dissection of the meat; not because they make good company but because the presence of meat forces them to (violently) share the same place, time, and fleeting resource. By contrast, the herbivores (vegetarians) are to be found in one herd, grazing together, each using their senses to ensure the group's collective security against suspicious movement.

The poolside experience suggests that it is not just humans that are blessed with the ability to measure and keep time. On the contrary, animals know when to come down to drink and when to depart the scene. It is not unusual to find young male baboons or impala sparring after having quenched their thirst. Having stood sentry, one of the primate patriarchs now descends to fill up its tank, as another takes over. To loiter around when the thirst is quenched is to expose oneself to the mercy of predators lurking in the water in front and the forests behind.[82]

Here, then, are codifications of space and time that are fleeting, visual forms presenting and rescuing themselves for/from observation, mobilities defined by a nutritional need—thirst and hunger. The internal workings of the digestive system as an energy production-consumption chain create the state of feeling thirsty. Thirst summons the body into movement to quench it. The imperative of satisfying a bodily need is always calculated into the measures the animal takes to secure its body not just from destructing from within (dehydration and starvation) but more so from physical harm by predators. This security of the body cannot function without energy, the driver of all kinetics.

The continued survival of species from generation to generation depends on the passage of genes; without mobilities and the quotidian production and expenditure of energy to drive movement, species become extinct. I pondered these thoughts when encountering two buffaloes along the Punda Maria–Crooks Corner road. This was undoubtedly a bachelor herd ostracized from the main herd following a losing battle with the incumbent or challenging patriarch. This is a sign of the hierarchy in the herds and the formation of patriarchy within the animal species. You are able to pass your genes only so long as your power, your strength of body, allows you to have sex with the females. If you cannot stand and fight to win or use your mobilities to migrate out of the reach of danger, you cannot sire any sons or daughters. Your genes die. So those two bachelor bulls show the connections between the masculinities that define human beings and those that define animals. They bring us closer to the humble reality that we are not too different from animals.

CONCLUSION

By now, we have arrived at a good point to restate and conclude on the role mobility plays in the making of animal meaning. An attention to the animal and what it is doing suggests that it is far from a mute presence in its own fate or that of space. The second part of the chapter addressed the question: how does the mobility of microbes shape human behavior toward the animal? The answer has revolved on how, by moving about inside the fluids or body of its carrier, the pathogen makes its carrier risky to different constituencies. Risk is not "socially neutral,"[83] but operates within and shapes other discourses and role players. Risk, in this specific instance, defined the conflicting moral judgments toward the use of specific technologies to kill the animal and the moods and times in which the animal was bad and justifying destruction one moment and good and worth preserving the next.[84] Each actor's response to the mobilities of microbe-carrying animals and insects could very well determine who was a good citizen and who was not.[85] The association of the animal with vermin is what stifled earlier efforts at conserving wild animals in the trans-Limpopo.

The last section of the chapter is a mobile ethnography of the animal as wildlife, or the good animal. By driving through the park with tape recorder in hand, making audio and visual notes of the human encounters with the animal, it was possible to see that the animal has a lot of say in defining humans to their own species. At the same time, the very existence and survival of an animal species depends entirely on the management of mobility. As southern Africa moves toward big transboundary conservancies that front nature so blatantly as something that is postnational, postpolitical, and anthropocentric, a selective dementia is reigning where nature is only thought of as good. Proponents and opponents of this utopian, Western vision have mobilized moral and legal arguments to make their cases. As the chapter has shown, there is a third way to examine thoroughly what is going on without recourse to social justification or moral judgment: we can examine how current and past mobilities of species and what they have carried on them have impacted what we erroneously call nature.

NOTES

1. Erich Fromm, *On Being Human* (New York: Continuum, 1997), 28.
2. Beyond, because Foucault meant it as the practice of modern states and their regulation of their subjects through "an explosion of numerous and diverse techniques for achieving the subjugations of bodies and the control of populations." See Michel Foucault, *The History of Sexuality*, vol. 1, *The Will to Knowledge* (London: Penguin, 1998), 120, 137.
3. For this reading of the "actor network" as a heuristic for tracing human-nonhuman interactions, see Bruno Latour and Steve Woolgar, *Laboratory Life: The Construction of Scientific Facts*, 2nd ed. (Princeton, N.J.: Princeton University Press, 1986); and Michel Callon, "Some Elements of a Sociology of Transition: Domestication of the Scallops and the Fishermen of Saint Brieuc Bay," in *Power, Action and Belief: A New Sociology of Knowledge?* ed. John Law (Boston: Routledge, 1986), 196–233.
4. As in Nigel Rothfels, ed., *Representing Animals* (Bloomington: Indiana University Press, 2002).
5. Africanists like Nancy Hunt, whose ethnographic and historical work examines the coproduction of a "colonial lexicon" by European colonists and inhabitants they found already living in Congo and colonized, have made the same point, albeit focusing on "the human," without accounting for "the nonhuman" as agent. See Nancy R. Hunt, *A Colonial Lexicon: Of Birth Ritual, Medicalization, and Mobility in the Congo* (Durham, N.C.: Duke University Press, 1999).
6. By the time that inquiry into science and technology as socially constructed gathered momentum in the 1960s-1980s, Ludwig Fleck had already published a seminal piece in 1935. See Ludwig Fleck, *The Genesis and Development of a Scientific Fact* (Chicago: University of Chicago Press, 1979). Parallel to emerging challenges to technological determinism in the 1960s was an interest in the history and philosophy of science that culminated in Thomas Kuhn's *The Structure of Scientific Revolutions* (Chicago: University of Chicago Press, 1962). The decisive "turn to technology" came in the 1980s with Donald Mackenzie and Judy Wajcman, eds., *The Social Shaping of Technology* (Buckingham, U.K.: Open University Press, 1985).
7. For some of the major works, see Sheila Jasanoff, *States of Knowledge: The Co-Production of Science and the Social Order* (London: Routledge, 2006); Hans Harbers, *Inside the Politics of Technology: Agency and Normativity in the Co-Production of Technology and Society* (Amsterdam: Amsterdam University Press, 2005); Gabrielle Hecht, *The Radiance of France: Nuclear Power and National Identity after World War II* (Cambridge, Mass.: MIT Press, 1998); Timothy Mitchell, *Rule of*

Experts: Egypt, Techno-Politics, Modernity (Berkeley: University of California Press, 2002); and Edmund P. Russell, *War and Nature: Fighting Humans and Insects with Chemicals from World War I to Silent Spring* (Cambridge: Cambridge University Press, 2001).

8. Callon, "Some Elements of a Sociology of Translation," 196–233. Among others, Latour's work has shaped the interest in actor networks. Apart from *Laboratory Life*, see also his *Science in Action: How to Follow Scientists and Engineers through Society* (Cambridge, Mass.: Harvard University Press, 1987); *The Pasteurization of France* (Cambridge, Mass.: Harvard University Press, 1988); *We Have Never Been Modern*, trans. Catherine Porter (Cambridge, Mass.: Harvard University Press, 1993); *Aramis, or the Love of Technology* (Cambridge, Mass.: Harvard University Press, 1996); *Pandora's Hope: Chapters on the Reality of Science Studies* (Cambridge, Mass.: Harvard University Press, 1999); *Politics of Nature: How to Bring the Sciences into Democracy*, trans. Catherine Porter (Cambridge, Mass.: Harvard University Press, 2004); with Peter Weibel, eds., *Making Things Public: Atmospheres of Democracy* (Cambridge, Mass.: MIT Press, 2005); and *Reassembling the Social: An Introduction to Actor-Network Theory* (New York: Oxford University Press, 2005).

9. For the most influential Donna Haraway texts on the animal, see *Primate Visions: Gender, Race, and Nature in the World of Modern Science* (New York: Routledge, 1989); *Simians, Cyborgs, and Women: The Reinvention of Nature* (New York: Routledge, and London: Free Association Books, 1991); *Modest_Witness@Second_Millennium.FemaleMan©Meets_OncoMouse™: Feminism and Technoscience* (New York: Routledge, 1997); *The Companion Species Manifesto: Dogs, People, and Significant Otherness* (Chicago: Prickly Paradigm Press, 2003); and *When Species Meet* (Minneapolis: University of Minnesota Press, 2008).

10. Harriet Ritvo, *The Animal Estate: The English and Other Creatures in the Victorian Age* (Cambridge, Mass.: Harvard University Press, 1987).

11. For the seminal works, see Peter Singer, *Animal Liberation: A New Ethics for Our Treatment of Animals* (New York: New York Review, 1990); Peter Singer and Lori Gruen, *Animal Liberation: A Graphic Guide* (London: Camden Press, 1987); and Peter Singer, "A Utilitarian Defense of Animal Liberation," in *Environmental Ethics*, ed. Louis Pojman (Stamford, Conn.: Wadsworth, 2001), 33–39. See also David Frum, *How We Got Here: The '70s* (New York: Basic Books, 2000); and J. M. Coetzee, *The Lives of Animals* (Princeton, N.J.: Princeton University Press, 2001).

12. Martin Heidegger, *Being and Time*, trans. John Macquarrie and Edward Robinson (New York: Harper and Row, 1962).

13. Emmanuel Levinas, *Of God Who Comes to Mind*, trans. Bettina Bergo (Stanford, Calif.: Stanford University Press, 1998).

14. Jacques Derrida, *Aporias: Dying—Awaiting (One Another At) the Limits of Truth*, trans. Thomas Dutoit (Stanford, Calif.: Stanford University Press, 1993).

15. Matthew Calarco, *Zoographies: The Question of the Animal from Heidegger to Derrida* (New York: Columbia University Press, 2008), 3.

16. The specific discussion Calarco refers to is found in Robert A. Wilson, ed., *Species: New Interdisciplinary Chapters* (Cambridge, Mass.: MIT Press, 1999).

17. Calarco, *Zoographies*, 5.

18. Calarco, *Zoographies*, 79, citing Giorgio Agamben, *The Open: Man and Animal*, trans. Kevin Attell (Stanford, Calif.: Stanford University Press, 2004).

19. Linda Kalof and Amy Fitzgerald, eds., *The Animals Reader: The Essential Classic and Contemporary Writings* (New York: Berg, 2007).

20. Jeffrey Moussaieff Masson and Susan McCarthy, "Grief, Sadness, and the Bones of Elephants," in Kalof and Fitzgerald, *Animals Reader*, 91–103.

21. Steve Baker, "What Does Becoming-Animal Look Like?" in Rothfels, *Representing Animals*, 68.

22. Among some of the latest works that examine the role nonhumans play in historical processes, see William Kelleher Storey, *Guns, Race, and Power in Colonial South Africa* (Cambridge: Cambridge University Press, 2008); Gabrielle Hecht, "Negotiating Global Nuclearities: Apartheid, Decolonization, and the Cold War in the Making of the IAEA," *Osiris* 21 (July 2006): 25–48; and Gabrielle Hecht, "Rupture-talk in the Nuclear Age: Conjugating Colonial Power in Africa," *Social Studies of Science* 32, nos. 5–6 (October–December 2002): 691–728, part of a special issue on "Postcolonial Technoscience," coedited with Warwick Anderson; Duana Fullwiley, "The Biologistical Construction of Race: 'Admixture' Technology and the New Genetic Medicine," *Social Studies of Science* 38, no. 5 (2008): 695–735; Stacey Langwick, "Articulate(d) Bodies: Traditional Medicine in a Tanzanian Hospital," *American Ethnologist* 35, no. 3 (2008): 428–439; and Stacey Langwick, "Devils, Parasites, and Fierce Needles: Healing and the Politics of Translation in Southern Tanzania," *Science, Technology & Human Values* 32, no. 1 (2007): 88–117.

23. J. R. McNeill. "Observations on the Nature and Culture of Environmental History," *History and Theory* (December 2003): 5–43. For the defining works, see Alfred Crosby, *Ecological Imperialism: The Biological Expansion of Europe, 900–1900* (Cambridge: Cambridge University Press, 1986).

24. See William Cronon, *Changes in the Land: Indians, Colonists, and the Ecology of New England* (New York: Hill & Wang, 2003).

25. William Beinart and Peter Coates, *Environment and History: The Taming of Nature in the USA and South Africa* (New York: Routledge, 1995).

26. Especially John McNeill, *The Atlantic Empires of France and Spain: Louisbourg and Havana, 1700–1763* (Chapel Hill: University of North Carolina Press, 1985); and John McNeill, *The Mountains of the Mediterranean World: An Environmental History* (New York: Cambridge University Press, 1992). McNeill is also part of an elite cast of scholars who write on world environmental history.

27. James C. McCann, *Maize and Grace: Africa's Encounter with a New World Crop, 1500–2000* (Cambridge, Mass.: Harvard University Press, 2005); James C. McCann, *Green Land, Brown Land, Black Land: An Environmental History of Africa* (Portsmouth, N.H.: Heinemann, 1999).

28. William L. Wolmer, *From Wilderness Vision to Farm Invasions: Conservation and Development in Zimbabwe's South-east Lowveld* (London: Routledge, 2007).

29. See Jane Carruthers, *The Kruger National Park: A Social and Political History* (Pietermaritzburg: University of Natal Press, 1995).

30. See Clapperton Mavhunga and Marja Spierenburg, "Transfrontier Talk, Cordon Politics: The Early History of the Great Limpopo Transfrontier Park in Southern Africa, 1925–1940," *Journal of Southern African Studies* 35, no. 3 (2009): 715–735.

31. Norma J. Kriger, *Zimbabwe's Guerrilla War: Peasant Voices* (Cambridge: Cambridge University Press, 1992).

32. Richard Waller, "'Clean' and 'Dirty': Cattle Disease and Control Policy in Colonial Kenya, 1900–40," *Journal of African History* 45 (2004): 45, 46–47; Richard Waller and Kathy Homewood, "Elders and Experts: Contesting Veterinary Knowledge in a Pastoral Community," in *Western Medicine as Contested Knowledge*, ed. Andrew Cunningham and Birdie Andrews (Manchester: Manchester University Press, 1997), 69–93.

33. William Wolmer, "Transboundary Conservation: The Politics of Ecological Integrity in the Great Limpopo Transfrontier Park," *Sustainable Livelihoods in Southern Africa Research Paper No. 3* (Brighton: Institute of Development Studies 2003), 19.

34. Conrad Steenkamp and J. Uhr, *The Makuleke Land Claim: Power Relations and Community-Based Natural Resource Management* (London: International Institute for Environment and Development, 2000); Maano F. Ramutsindela, "The Perfect Way to Ending a Painful Past? Makuleke Land Deal in South Africa," *Geoforum* 32 (2002): 15–24.

35. Terence Osborne Ranger, *Voices from the Rocks: Nature, Culture & History in the Matopos Hills of Zimbabwe* (Bloomington: Indiana University Press, 1999).

36. Isak Niehaus, "Ethnicity and the Boundaries of Belonging: Reconfiguring Shangaan Identity in the South African Lowveld," *African Affairs* 101 (2002): 557.

37. Niehaus, "Ethnicity and the Boundaries of Belonging," 558.

38. Latour, *Pandora's Hope*, 176–179.

39. Much of this effort, inspired by the mentorship of William Beinart, has come from two outstanding young scholars, Daniel Gilfoyle and Karen Brown. See Karen Brown and Daniel Gilfoyle, eds., *Healing the Herds: Disease, Livestock Economies and the Globalization of Veterinary Medicine* (Athens: Ohio University Press, 2010); Karen Brown, "From Ubombo to Mkhuzi: Disease, Colonial Science and the Control of *Nagana* (Livestock Trypanosomiasis) in Zululand, South Africa, c. 1894–1955," *Journal of the History of Medicine and Allied Sciences* 63, no. 3 (July 2008): 285–322; Karen Brown, "Frontiers of Disease: Human Desire and Environmental Realities in the Rearing of Horses in 19th and 20th Century South Africa," *African Historical Review* 40, no. 1 (2008): 30–57; Karen Brown, "Poisonous Plants, Pastoral Knowledge and Perceptions of Environmental Change in South Africa, c. 1880–1940," *Environment and History* 13, no. 3 (2007): 307–332; Karen Brown, "Tropical Medicine and Animal Diseases: Onderstepoort and the Development of Veterinary Science in South Africa 1908–1950," *Journal of Southern African Studies* 31, no. 3 (September 2005): 513–529; Karen Brown, "Trees, Forests and Communities: Some Historiographical Approaches to Environmental History on Africa," *Area* 35, no. 4 (December 2003): 343–356; Karen Brown, "Agriculture in the Natural World: Progressivism, Conservation and the State. The Case of the Cape Colony in the Late 19th and Early 20th Centuries," *Kronos Special Edition on Environmental History* 29 (November 2003): 109–138; Karen Brown, "Political Entomology: The Insectile Challenge to Agricultural Development in the Cape Colony 1895–1910," *Journal of Southern African Studies* 29, no. 2 (June 2003): 529–549; Karen Brown, "Cultural Constructions of the Wild: The Rhetoric and Practice of Wildlife Conservation in the Cape Colony at the Turn of the Twentieth Century," *South African Historical Journal* 47 (November 2002): 75–95; Karen Brown, "The Conservation and Utilization of the Natural World: Silviculture in the Cape Colony circa 1902–1910," *Environment and History* 7, no. 4 (November 2001): 427–447; Karen Brown and Daniel Gilfoyle, eds., "Introduction: Livestock Diseases and Veterinary Science in Southern Africa" special edition on environment and livestock diseases in *South African Historical Journal* 58 (2007): 2–16; William Beinart, Karen Brown, and Daniel Gilfoyle, "Experts and Expertise in Colonial Africa Reassessed: Colonial Science and the Interpenetration of Knowledge," *African Affairs* 108, no. 432 (2009): 413–433. See also Daniel Gilfoyle, "Veterinary Immunology as Colonial Science: Method and Quantification in the Investigation of Horsesickness in South Africa, c. 1905–1945," *Journal of the History of Medicine and Allied Sciences* 61, no. 1 (January 2006): 26–65.

40. Maryinez Lyons, "African Sleeping Sickness: An Historical Review," *International Journal of STD & AIDS (Supplement 1)* 2 (1991): 20–25.

41. Maryinez Lyons, "Sleeping Sickness Epidemics and Public Health in the Belgian Congo," in *Imperial Medicine and Indigenous Societies*, ed. David Arnold (Manchester: Manchester University Press, 1988).

42. James Giblin, "Trypanosomiasis Control in African History: An Evaded Issue?" *Journal of African History* 31, no. 1 (1990): 59–80; Maryinez Lyons, *The Colonial Disease: A Social History of Sleeping Sickness in Northern Zaire, 1900–1940* (Cambridge: Cambridge University Press, 1992); John J. McKelvey, *Man against Tsetse: Struggle for Africa* (Ithaca, N.Y.: Cornell University Press, 1973); John M. MacKenzie, "Experts and Amateurs: Tsetse, Nagana and Sleeping Sickness in East and Central Africa," in *Imperialism and the Natural World*, ed. John M. MacKenzie (Manchester: Manchester University Press, 1990), 187–212.

43. Maureen Malowany, "Unfinished Agendas: Writing the History of Medicine of Sub-Saharan Africa," *African Affairs* 99 (2000): 334; J. Andrew Mendelsohn, "From Eradication to Equilibrium: How Epidemics Became Complex after World War I," in *Greater Than the Parts: Holism in Bio-medicine 1920–1950*, ed. Christopher Lawrence and George Weisz (Oxford: Oxford University Press, 1998), 304.

44. Village Innovations Museum (VIM): Ecology: Elmon Chauke, Attorney Hlongwane, and Benes Maluleke, "Interviews with Ex-Poachers," Makuleke Village, South Africa, June 30–July 10, 2009, Video Tape #3.

45. For this and perhaps the freshest theorizing on the topic from southern Africa, see Jessica Dubow, "'From a View on the World to a Point of View in It': Rethinking Sight, Space and the Colonial Subject," *Interventions* 2, no. 1 (2000): 89–90; Jessica Dubow, "The Mobility of Thought: Reflections on Blanchot and Benjamin," *Interventions: The International Journal of Postcolonial Studies* 6, no. 2 (2004): 216–228; Jessica Dubow, "Out of Place and Other Than Optical: Walter Benjamin and the Geography of Critical Thought," *Journal of Visual Culture* 3, no. 3 (2004): 259–274; and Jessica Dubow, "Minima Moralia: Or the Negative Dialectics of Exile," in *Imaginary Coordinates*, ed. R. Rosen (Chicago: Spertus Institute of Jewish Studies, 2008).

46. Clapperton Mavhunga, "Big Game Hunters, Bacteriologists, and Tsetse Fly Entomology in Colonial South-East Africa: The Selous-Austen Debate Revisited, 1905–1940s," *ICON* 12 (2006): 75–117.

47. Helge Kjekshus, *Ecology Control and Economic Development in East African History*, 2nd ed. (London: James Currey, 1996).

48. Scholarship on the rinderpest, including that which looks at what people were doing to rinderpest but not what the rinderpest was doing to them and to define livestock as vermin, is still sparse. For more serious recent work, see Pule Phoofolo, "Epidemics and Revolutions: The Rinderpest Epidemic in Late Nineteenth-Century Southern Africa," *Past & Present* 138, no. 1 (February 1993): 112–143; and Pule Phoofolo, "Face to Face with Famine: The BaSotho and the Rinderpest, 1897–1899," *Journal of Southern African Studies* 29, no. 2 (June 2003): 503–527.

49. C. F. M. Swynnerton, "An Examination of the Tsetse Problem in North Mossurise, Portuguese East Africa," *Bulletin of Entomological Research* 11 (1921): 315.

50. John Ford, *The Role of Trypanosomiases in African Ecology* (Oxford: Clarendon Press, 1971), 334.

51. Anon., "The 'Rinderpest': What It Is with Symptoms and Causes: An Interview with Dr. Hutcheon," *African Review* (May 23, 1896): 1027.

52. Anon., "Interview with Mr. Lionel Decle, the African Explorer: Interesting Facts about the Matabele," *African Review* (May 9, 1896): 917; Anon., "The Situation in South Africa: Origin of the Matabeli Revolt," *African Review* (April 18, 1896): 752.

53. Anon., "The Hon. John Scott Montagu, MP, on the Matabeli Rising and Its Causes," *African Review* (June 6, 1896): 1121.

54. Anon., "Interview with Mr. Lionel Decle," 917.

55. Anon., "The Cattle Plague in South Africa: A Very Serious Outlook," *African Review* (June 6, 1896), n.p.

56. Anon., "Cattle Plague in South Africa."

57. Allan Wright, *Valley of the Ironwoods: A Personal Record of Ten Years Served as District Commissioner in Rhodesia's Largest Administrative Area, Nuanetsi, in the South-Eastern Lowveld* (Cape Town: T. V. Bulpin, 1972), 329.

58. See Martin Murray, "Blackbirding at Crooks' Corner: Illicit Labor Recruiting in the Northeastern Transvaal, 1910–1940," *Journal of Southern African Studies* 21, no. 3 (1995); VIM Royal Collections: Royal Guided Tour to Old Makuleke, "Videotape of Conversation with Chief Makuleke (Joas Pahlela) at Crooks' Corner, on the South Bank of the Limpopo River" (June 2008).

59. I deal with the state's deference to lethal mobility (armed police patrols) against these "white bandits" in "The Mobile Workshop: Mobility, Technology, and Human-Animal Interaction in Gonarezhou (National Park), 1850–Present" (Ph.D. diss., University of Michigan, 2008), chapters 4 and 5.

60. Mavhunga and Spierenburg, "Cordon Politics."

61. There is no reason why what Jacques Semelin describes as the human source of and imperative to massacre other humans in *Purify and Destroy: The Political Uses of Massacre and Genocide* (New York: Columbia University Press, 2007) cannot be extended to examining similar human impulses against other species.

62. See National Archives of Zimbabwe (NAZ) S3106/11/1/1–9 Sabi Valley 1947–56.

63. NAZ, S1215/1880/1 Nuanetsi Ranch—Boundary Fencing 1935–43; S3106/11/1/3 Sabi Valley 1940–4; S2136/58779/66 Fencing Prohibited Area, South Eastern Districts 1937.

64. NAZ, S2376/S58779/49 African Coast Fever: Fencing of Portuguese Border on Sabi River 1935–42: Chief Veterinary Officer to Secretary Agriculture, October 7, 1935; Anon., "Combating Foot and Mouth Disease—Drastic Slaughter Policy Adopted," *African World* (August 18, 1934): 181; Anon., "Inter-State Veterinary Conference Proposed—to Investigate Foot-and-Mouth Question," *African World* (November 3, 1934): 101.

65. NAZ S914/12 Game Reserves in Southern Rhodesia 1933–35.

66. The chief architect of the game reserve idea, who seemed to be in the pocket of the commercial and naturalists interests, was R. D. Gilchrist, the secretary for commerce, transport, and public works during the 1930s. See NAZ, S914/12/1D, Chipinda Pools—Proposed Reserve 1933–4; NAZ, S1194/1645/3/1, Proposed Game Reserves—Chipinda Pools and Gwanda 1932–35.

67. Luise White, "Tsetse Visions: Narratives of Blood and Bugs in Colonial Northern Rhodesia, 1931–9," *Journal of African History* 36 (1995): 223, 232.

68. VIM: Ecology: Clapperton Mavhunga, Personal Files: Mobilities, "Audio Record of Seven Days of 'Mobile Ethnography' inside the Whole of Kruger National Park, Entering at Malelane Gate and Exiting at Punda Maria and Back" (June 30–July 8, 2009): Clip on "The Gate."

69. VIM: Ecology: Clapperton Mavhunga, Personal Files: Mobilities, "Audio Record of Seven Days of 'Mobile Ethnography'": Clip on "Cars Stopping at Attraction."

70. VIM: Ecology: Clapperton Mavhunga, Personal Files: Mobilities, "Audio Record of Seven Days of 'Mobile Ethnography'": Clip on "The Attraction: Seeing through Preconceptions and Meaning-Mapping."

71. VIM: Ecology: Clapperton Mavhunga, Personal Files: Mobilities, "Audio Record of Seven Days of 'Mobile Ethnography'": Clip on "The Attraction."

72. VIM: Ecology: Clapperton Mavhunga, Personal Files: Mobilities, "Audio Record of Seven Days of 'Mobile Ethnography'": Clip on "Bagging the Attraction: Shooting without Killing."

73. VIM: Performing Arts—GG Drama Club: "Drama and the Capturing of the 1969 Eviction."

74. VIM: Ecology: Clapperton Mavhunga, Personal Files: Mobilities, "Audio Record of Seven Days of 'Mobile Ethnography'": Clip on "Observations of the Elephant's Environmental Destruction of Trees in the Singita Area, Along Gudzani Road."

75. For detailed ethnography on Shangane beliefs on nature, see Mavhunga, "Mobile Workshop," this volume.

76. VIM: Ecology: Clapperton Mavhunga, Personal Files: Mobilities, "Audio Record of Seven Days of 'Mobile Ethnography'": "Conversations with Elmon Chauke, Attorney Hlongwane, Benes Maluleke, and Denise Ortiz While Driving to/back from Old Makuleke, 16 June 2009."

77. VIM: Ecology: Clapperton Mavhunga, Personal Files: Mobilities, "Audio Record of Seven Days of 'Mobile Ethnography'": "Conversations."

78. VIM: Ecology: Clapperton Mavhunga, Personal Files: Mobilities, "Audio Record of Seven Days of 'Mobile Ethnography'": "Conversations."

79. "Battle at Kruger," http://www.youtube.com/watch?v=LU8DDYz68kM.

80. "Battle at Kruger."

81. "Battle at Kruger."

82. VIIL Ecology: Clapperton Mavhunga, Personal Files: Mobilities, "Audio Record of Seven Days of 'Mobile Ethnography' inside the Whole of Kruger National Park, Entering at Malelane Gate and Exiting at Punda Maria and Back" (June 30–July 8, 2009): Clip on "Poolside Observations at Lower Sabi Bridge."

83. A. Molella and J. Bedi, eds., *Inventing for the Environment* (Cambridge, Mass.: MIT Press, 2003), xii.

84. Dale Rose and Stuart Blume, "Citizens as Users of Technology: An Exploratory Study of Vaccines and Vaccination," in *How Users Matter*, ed. Nelly Oudshoorn and Trevor Pinch (Cambridge, Mass.: MIT Press), 103.

85. Richard White, "Tempered Dreams," in Molella and Bedi, *Inventing for the Environment*, 3–10.

Cannibalism, Consumption, and Kinship in Animal Studies

ANALÍA VILLAGRA

HIS POWERFUL BODY IS EERILY INERT. BURLY ARMS, BARREL-CHEST FLECKED WITH the silvery gray of age and leadership, the strong face completely still as the villagers soberly carry his body out of the forest and through the sunlit field. His was no accidental death, but a murder, a political killing at the center of civil strife in a national park in the Democratic Republic of Congo. The victim described here is the silverback male of the Virunga gorilla troop, one of seven gorillas killed "in cold blood" in July 2007.[1] The Virunga forest has been in the news before, famous as the area in which Dian Fossey worked. Today, as then, countries torn apart by poverty, guerrilla warfare, and desperation surround the park. In such conditions nature reserves are often at odds with the local population, who might be more inclined to consume the rare animals contained within as bush-meat rather than consuming them as icons of an impressive natural heritage as ecotourists might. The consumption of primates, and other endangered species, as bush-meat is certainly a problem (which will be addressed later in this chapter), but this recent Virunga case is exceptional in that it is not about bush-meat. When photographer Brent Stirton's picture of the slain gorilla entered the international news circuit, the disturbing image was accompanied by language equally disturbing, not only for the visceral imagery but also for its unexpected application of human terms to the death of a gorilla. These gorillas were not killed for meat, nor were they hunted or poached. Rather, they were *murdered*, "shot in the back of the head, execution style."[2]

We are accustomed to hearing about violence, murder, and political intrigue; people are murdered and leaders assassinated with numbing frequency.[3] It is not the fact of murder in and of itself that makes this story so sensational. But multiple dimensions of an *animal's* murder are distressing, thought-provoking, unsettling. The description of seven gorillas "murdered" as part of an ongoing political power struggle over national park land grants the animals an uncomfortable human subjectivity. Creatures who were once the sacred objects of our conservation efforts now have a larger, more personal stake in the politics that protect or condemn them on their forest reserve; put simply, they have moved from objects to subjects.

Although he rejects the idea that an animal can be murdered, Jacques Derrida finds that the denial of subjecthood and the lumping of all nonhumans into the absurdly broad, undifferentiated category of "animals" amounts to nothing less than "violence against animals," a violence that is at the root of all human violence against nonhumans.[4] The description of the physical attack on the Virunga gorillas as a "murder" directly challenges this imbalance, described by Derrida as the "denegation of murder."[5] To suggest that the group was "poached" rather than murdered would extend the brutality committed against the gorillas to include both the physical and the philosophical, a denial of the right to life and the right to subjecthood. The denegation asserts that animals are not murdered, but rather put to death.[6] Murdered gorillas, and the attendant suggestion of the gorilla-as-subject, go a long way toward correcting this long-standing human hostility toward nonhumans. However, Derrida finds this violence is unavoidable, even by vegetarians or other similarly ethically conscious people, for, as Matthew Calarco summarizes, "in order to speak and think about or relate to the Other, the Other must—to some extent—be appropriated and violated, even if only symbolically."[7] In this frenetic and violent moment in the Virunga forest, our relationship to the gorilla as a taxonomic kin, the consumption of gorillas as potential bush-meat, and the sudden subject-hood of particular gorillas come together. From this jumble, from this confusion of categories, how are we to relate to the gorilla? Derrida does "not believe in the existence of the non-carnivore in general."[8] The nature of man's gaze upon this hydra-headed concept of "the animal" is a vicious, carnivorous gaze, and "in a more or less refined, subtle form, a certain cannibalism remains unsurpassable."[9] The remainder of the chapter explores this problem by situating it within the unlikely, but analytically informative, context alluded to by Derrida, that of the cannibal.

Cannibalism is viewed as the most abhorrent of transgressions. Accusations of cannibalism have been levied against numerous groups of people as a means with which to construct them as most fantastically other, as less than human. What could be more barbaric than consuming one's own kin? The swift and visceral reaction to the idea of the cannibal becomes complicated when it is placed alongside the burgeoning field of animal studies where we are asked to consider animals as beings worthy of historical, social, and moral consideration. What do we become when animals are consumed as food and loved as kin? At this nexus of animal consumption and animal kinship we find ourselves faced with the curious theoretical specter of cannibalism. Like the murdered gorilla, the cannibal folds together categories that we held as separate. By examining multiple dimensions of this cannibal problem I hope to contribute to the larger, broader question of how we relate to animals, of how the figure of the cannibal adds intrigue to this question, and finally seek out the persistent limitations to the contentious pursuit of animal kinship. What is the meaning of such kinship if it can be so easily consumed? The roots to this question are firmly planted in our primate cousins.

THE CONSUMPTION OF PRIMATES

While not a story of bush-meat, the Virunga gorillas are certainly situated within the narrative of bush-meat. A dead gorilla in an African park immediately suggests yet another victim of the bush-meat trade, a practice that numerous other articles and photo essays have documented.[10] The name suggests primitivism. The "bush" is not a forest or a grassland or the territory of a nation, but rather a frontier, the edge of civilization. "Bush-meat" thus evokes a kind of raw carnage that is distinctly inhuman. Photos of bush-meat are not of cooked or butchered meat, safely unrecognizable like the neatly packaged ground beef or sirloin that one finds in an American grocery store. An oddly human arm tossed onto a burning cook fire or a startled head, tongue lolling to the side: these are the trademarks of so savage a meal as one would find deep in the African bush, far from civilized, abstract, and sanitary meat. Never mind that the pert pink packages we are familiar with hide a messy and barbaric factory farming industry;

civilization demands that mask.[11] Nonindustrial meat products—carcasses for butchering out in plain view, parts that are recognizable, enough blood to suggest an actual, living creature—are shameless in their refusal to participate in the Victorian modesty that would hide meat's messy origins.[12] Primate bush-meat is further implicated. Not only is it flagrantly the product of a living creature, that creature is uncomfortably similar to human. The arm on the campfire has five fingers and a familiar musculature; the startled head looks startled because it shares expressions we have seen among our kind. The transgression illustrated by the consumed ape is double, thus the reaction to it is more violent.

No properly civilized person would deign to make a meal of such a humanlike creature; this meal hints of the flouting of the cannibalism taboo. Much like accusations of cannibalistic practices among certain cultural groups (which will be discussed in detail below), those who engage in primate-eating might automatically equate to barbaric subhumans. Thus, evidence of cultures that do not share our hesitation to eat primate meat must be carefully approached and scrutinized for implicit primitivizing. We would be wise to remember Derrida's admonition against embracing such imprecise labels as "animal," or in this case "cannibals" or "primitives." Such categories fail to account for discriminating tastes and well-differentiated practices that nonetheless receive the same label from outsiders. As is discussed below, all cannibalism is not the same. The consumption of primates as well is by no means indiscriminant. Rather than a free-for-all, as stereotypic ideas about the "wild man" or "untouched primitive" might suggest, primate consumption is subject to species-differentiated taboos and preferences, as are any other less controversial food products. In an example of the plurality of indigenous views of not just animals but nonhuman primates in particular, anthropologist Axel Köhler notes the differences between the neighboring Bantu and Baka of central Africa in terms of their sense of identification with different groups of great apes. Though they have a reputation as formidable hunters, most Baka "deem the great apes to be too person-like, too close to human beings both in shape and in behavior" to hunt,[13] while "the idea of eating a close relative of their own species, however, does not seem to bother many Bantu."[14] Practices of eating animals are also often deeply embedded into cosmological schemes. Such beliefs render eating "not an act that merely satisfies hunger; it also has the transformational power to make another sacred."[15] The transformational power of eating suggests that the primitivizing implications of bush-meat stem from the lack of critical focus on the many ways that primates are consumed.

The consumption of primates is not a practice that is exclusively, or even predominantly, the field of the primitivized human being. Primates consume and are consumed by other animals. But primates are our taxonomical kin. We share a suite of traits with the other members of our Order; primates, like us, should lord over their respective food chains. However, the natural politics of eating and being eaten reveals an all-too-animal position for our taxonomic fellows. The relationship of the vervet monkey and the African crowned eagle presents an interesting case. Not only does the eagle eat small mammals (an expected move for a raptor), it eats primarily small monkeys, particularly the vervet.[16] Our understanding of primates as among the "higher" mammals and sharing an Order with us raises our expectations for their dignified and dominating existence in the natural world. Granted, many animals may occasionally feed on monkeys—big cats, raptors, large reptiles—but for the most part large omnivorous primates may eat smaller mammals and small omnivorous primates feed on smaller reptiles and bugs; the primates always land near the top of the food chain or web to which they belong. Not so with the case of the crowned eagle and the vervet. In this instance the vervets occupy a second-place position in the food hierarchy, making up an enormous percentage of the eagle's diet (over 80 percent as some studies have suggested.)[17] How did our relatives end up in such an unfortunate, ignoble place? More important, how does this relate to the specter of the cannibal?

The predation of our genetic kin is unfortunate, but despite the closeness of our DNA,[18] the vervet does not inspire an instant identification with human beings. The chimpanzee is a different matter. The chimpanzee is genetically, physiologically, behaviorally, and socially more similar to us than any other living animal.[19] Popular science has impressed upon us the fact that the chimp is our closest living relative, while popular culture has reflected our enduring struggle to accept this connection; the chimp is an animal possessed of culture, technology, and social politics that make intuitive sense to us. And just like (many of) us, chimps hunt and eat other animals. In fact, chimps are well known for the organization of expertly coordinated hunting parties. We can easily see ourselves, and our evolutionary beginnings, in a chimp eating a bird, a fish, or even a small reptile, but chimps frequently hunt other primates, such as the colobus monkey. Given their similarities to us, it makes us uncomfortable to imagine consumption practices that begin to push against the taboo of cannibalistic consumption. In a chilling scene in David Attenborough's popular documentary series *The Life of Mammals*,[20] the joyful and excited hoots of the chimpanzee hunting party echo through the forest as it chases a terrified colobus monkey clutching her wailing baby. We have a sense that the chimps have crossed some line, breached some kind of unspoken contract of our taxonomic order, but the chimps give no thought to eating this other creature who shares its large brain size, its postorbital bars, its tactile pads complete with fingerprints. An illustration of a similar encounter graces the cover of Donna Haraway's *Primate Visions*,[21] the human hand inside the structurally similar hand of the ape showing a simple yet profound moment in our burgeoning sense of interspecies kinship. The humanlike features of nonhuman primates that capture researchers and zoo-goers alike strike a deep chord and recall us to a family tree with roots millions of years deep. But chimps do not exhibit the same sense of wonder that we do, this sympathetic vision of Order. If chimps do not feel the sense of kinship that we feel with them, is there such a thing as interspecies kinship? Perhaps it is merely a vast quixotic fiction, a sensibility confined to the romantically inclined academic. If no natural repulsion keeps the chimp from eating monkeys, then is there any hope for interspecies kinship escaping the dark evolutionary impulse toward cannibalism?

THE HISTORICAL CANNIBAL

The relatively recent notion of interspecies kinship has the formidable task of contending with the far more established and historically rooted intellectual tradition surrounding the cannibal. The figure of the cannibal shares its genesis with the birth of humankind. The dark mythology of our species is one of insatiable bloodlust and hunting instincts, in which "predation was one of the key forces in the process of humanization."[22] Within that story of triumph over any and all, adversity hides, tucked into its dark recesses, the skulking possibility of a cannibal past. After all, anatomically modern humans coexisted for a time with Neanderthals, who dabbled in art and religion before our ancestors. What became of our hominid fellows?[23] What kind of violence were we capable of? We cannot simply brush aside the untoward suggestion of such observations as mere suspicion for, as Cartmill reminds us, "good myths embody big truths."[24] Moving beyond the genesis of bloodthirsty man in prehistoric times, the image of the cannibal becomes even more vivid in historical time (see figure 1).

Figure 1. Theodo de Bry, *Cannibalism*. (Source: Wellcome L0005638)

History is rife with allegations of cannibals. Throughout history many peoples have been accused by others of practicing cannibalism, such as the Christians by the Romans and the Jews by the Christians, evidence of a persistent ethnocentrism that William Arens calls a syndrome of "others but not us."[25] Travelers to strange lands return home with reports of men who eat other men, living far off on isolated islands, in distant corners of the earth. While the authors of such reports nearly always claim to have seen "with their own eyes" these grotesque foreigners consuming the flesh of other human beings, the likelihood of this witness feels slight, not the product of outright lies, but rather evidence of a pervasive conviction in the widespread existence of cannibalism, practiced by protomen at the edge of civilization. Such historical accusations are swiftly brushed aside by experts who find evidence of their own cannibal past to be conjectural and inadequate, while accepting similar evidence of other, foreign cultures.[26] More contemporary scholars are far more likely to accept the suggestion of cannibalism among tribes in Africa than of their own Christian forebears.

The suggestion that a group of people consumed human flesh has always been a highly political move. Such accusations reduced the people encountered during colonial conquest to animals, and in some cases justified their treatment as such. The Spanish conquistadors used the practice of cannibalism as a shining example of the subhumanity of the New World Natives. Queen Isabella forbade the enslavement of Native peoples, excepting those who practiced anthropophagy, an exception that led to numerous indigenous groups becoming conveniently labeled as cannibals.[27] Allegations and accusations of cannibalism by European explorers constructed the Native peoples as more animal than human, closer in behavior to the exotic nonhuman fauna that they encountered, and with this construction easing any potential moral qualms about the brutality of the conquest. Steeped in the same kind of mythic folklore and Western intellectual history that swiftly and uncompromisingly made sense of the story of man as a violent, hunting ape,[28] even anthropologists are taken to task for their role in perpetuating the seductive myth of the primitive cannibal through unwarranted emphasis and selective credulity of unconfirmed (and unlikely) stories.

OF CANNIBALS AND KIN

The political and social reverberations of the accusation of cannibalism, and the irresponsible swiftness of so many, intellectuals and laypersons alike, to embrace thin evidence of cannibalism leads Arens to reject the existence of such a practice in any place at any time.[29] So sweeping a dismissal, while making a valid point about the terrifying ease with which the slightest hint of such a practice leads to dehumanizing unfamiliar people, rejects reasonable evidence of ritual and metaphoric cannibalism and precludes careful attention to these important cultural practices. Admittedly, the contentious history of cannibal accusation begs the question of why the idea of cannibalism persists as the most abhorrent of transgressions, a kind of pancultural gloss for the inhuman Other. What is it about the cannibal that strikes such an uneasy note? With its close ties to kinship, suggestions of animality and othering, this question has relevance for the work proposed by animal studies. Setting aside the debates about the historical frequency of cannibalism, the *idea* of the cannibal presents the most interesting problem for animal studies. Animal studies asks us to sidle up next to our animal kin, but many of these potential kin are also potential food, thus presenting an unresolved tension between cannibal and kin. Before we comfortably accept animals as kin we must confront the problem of the animals' edibility. Either we consume our kin and make cannibals of ourselves or we deny their kinship at the moment of consumption. The former breaches historical cultural taboos, and the latter makes a mockery of the sacred relations of kin. I would like to argue for a more challenging vision of kinship that would allow for the consumption of fellow animals not in the absence of or in spite of bonds of kinship, but rather because of them.

The question of animal kinship returns us to the previous discussion of chimps and colobus monkeys. We argue for a sense of kinship with the chimp based on physical, social, and genetic similarities, but does the chimp feel kinship with us? And a more difficult question, can we argue for a sense of kinship between the chimp and the colobus? If chimps and colobus cannot be kin then it is only our arrogant imagination that presumes that animals can relate to us as they cannot relate to one another.[30] That is not kinship, but kingship, Adam once again naming the beasts in the Garden. It is in response to this problem that the figure of the cannibal emerges and points the way toward new possibilities for thinking about kinship.

An expansive view in which "cannibal consumption is any devouring (literal or symbolic) of the other in its (raw) condition"[31] helps us to imagine how cannibalistic practice produces and adjusts kinship relationships. Consumption of a body, human or nonhuman, refigures the relationship between man and other animals as well as between men. And, although food is a basic human need, varying practices of eating become "a means of creating cultural difference,"[32] in which "eating produced an alliance among those who ate together and separated those who were, potentially, food for one another."[33] This suggests that the thoughtless eating of people is not an acceptable act, even among societies with an established practice of ritual anthropophagy. While a number of groups seem to have at least a tradition of reference to anthropophagic practice,[34] their views on flesh-eating may actually share similar values of human flesh with the Western taboo against cannibalism. The cosmological view of some indigenous South American peoples suggests that no creature, animal or human, envisions itself in the untoward act of consuming human flesh. One of the first descriptions of Brazil to enter the historical record is German seaman Hans Staden's account of his captivity among the cannibalistic Natives (see figure 2). Staden quotes his captor, who says that when he eats another man "I am a tiger [jaguar]; it tastes well."[35] In other words, he is not a man eating another man, but rather a member of the animal kingdom, a respected forest carnivore, performing an ultimately natural act.[36] While there may have been a touch of facetiousness in the comment of Staden's captor, the contemporary Amazonian cosmology of the Wari' people suggests the same reversal of identification; the jaguar "sees himself as a man walking upright and carrying a bow and arrow, for his claws are the bow and his teeth are the arrows. When this jaguar-hunter meets a person, to the jaguar's eyes, the person looks like a jaguar, so he shoots it."[37]

A far cry from the trope of the primitive man indiscriminately gnawing on the limbs of his weaker fellows, the richly spiritual practice of cannibalism takes careful measure of who is eaten, when, where, and how. Rarely, if ever, inspired by caloric need, cannibalism has little to do with a "taste" for human flesh and instead represents a process of transformation and refiguring of relationships. The blurry boundary between human and nonhuman animal does not begin and end with the actual consumption of flesh. Hans Staden's treatment by his captors indicates the expression of fluid categories. Though destined for ceremonial anthropophagy, Staden describes himself as a pet[38] and another slave as part of the lineage of his captors.[39] Not only Staden but other European and non-European captives were well treated and occasionally even given wives with whom they produced children, until the time was right to ritually consume them.[40] Animal, and animalized human, bodies negotiate the border of friend and food.

The eating of human flesh is one of a number of Native practices that incorporates indigenous peoples into the animal kingdom. Other Native practices—including the raising of baby monkeys by young girls in the Amazon as a means to prepare them for motherhood while expanding the family group across species borders,[41] and narrative traditions featuring kingdoms of dolphin-men who delight in entering the human world to create sexual mischief[42]—also emphasize the permeability of the human-animal boundary. Nature and culture are folded into one another. While nothing quite as dramatic as the cannibalism taboo is breached in these other examples of human-animal interaction, animal and human bodies traverse boundwaries of nurturing and sexual behavior in such a way as to open up the possibility

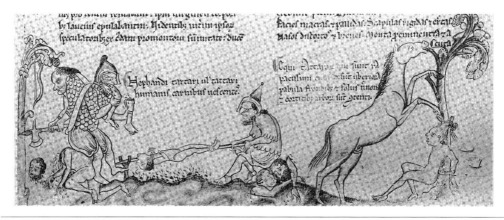

Figure 2. A Tartar Cannibal Feast, by Matthew Paris. (Source: *The Art of Matthew Paris in the Chronica Majora*, by Suzanne Lewis. Berkeley: University of California Press in collaboration with Corpus Christi College, Cambridge, 1987, Figure 180)

for that final, abhorrent transgression of eating the flesh of like creatures. This might inspire discomfort for Western audiences whose intellectual legacies are deeply influenced by Christian nature-culture distinctions.

BECOMING CANNIBAL

The act of cannibalism deserves closer attention for a more nuanced understanding of the profound, transformative work that it performs in the service of entwining the human with the animal. Even the most outlandish accusations of cannibalism generally do not suggest that human flesh makes up a regular part of any society's diet; rather, such a practice is associated with ritual and specific occasions. For example, Wari' mortuary cannibalism serves as the first step in the process of the human spirit reappearing on Earth in the form of a white-lipped peccary.[43] Bodies of family members are consumed to hasten the release of the spirit, a spirit that may return to the land of the living in the form of this important food source; the recently dead, their memories of their families still fresh, may even deploy some of their new peccary kin to offer themselves as food for the surviving human family.[44] Clearly, cannibalism is not a wild act of aggression, but rather a deeply social process that absorbs eating and eaten humans into the animal kingdom. The reinsertion of human sociality into the wider natural world requires the employment of more animal terms of relation. Animals relate to one another in complex, fluid food webs. Although an animal might consume another animal, that predator might in turn become the prey of another. Animals do not have to take on a permanent identity as eating or eaten animals; they may weave back and forth through a series of relationships and thus a series of different identities.

By carefully considering the cannibal as more than just a perversity or anomaly, we can begin to perceive of human beings in such fluid, relational terms. Taking the idea of interspecies kinship seriously demands that kind of flexibility with the human position in the natural world. Among the Araweté, an Amazonian society with an established narrative of divine cannibalism, "there is no taxon for 'animal,'" rather terms to describe animals describe human-animal relationships such as "for eating" or "potential pets."[45] However, despite the promise of relational categories, in this cosmological construction animals are made, versus human beings and spirits who "simply exist."[46] The employment of relational, as opposed to absolute, terms to describe the nature and identity of an animal is instructive, but still leaves kinship on shaky foundations. Kinship cannot be constructed if the human being possesses an untouchable dominance, a position of privilege that we cannot even begin to question. For a more rigorous and enduring model of kinship, we must confront the idea that as the kin of other animals, when we consume them we may become the cannibals we have so feared. We must also open ourselves to the possibility that our fears have been misguided. "Taking cannibals seriously" entails releasing ourselves from the cultural mythology of the ruthless, inhuman killer (the Hannibal Lecter mythos) and acknowledging the profoundly transformative aspect of cannibalism.

Our "becoming cannibal" is not so tragic. There is an honesty in accepting a cannibal identity, and cannibal consumption is certainly more productive and noble than the way we currently consume animals. Unlike the destructive and demeaning consumption of animals as commodified pieces of factory-farmed meat in a global capitalist marketplace, cannibal consumption seeks to incorporate rather than alienate. Not only the body destined for consumption is transformed; the body that does the consuming is also reshaped by this transaction. Cannibal consumption may come to represent a union between two bodies rather than expressing domination of one over another. As animals move from the object world to the subject world, they challenge the assumption that they are destined for eternal human domination. The most frightening thing about taking animals seriously may be that we must begin to release our grip on the sacred right of human privilege, to accept occasional domination by animal subjectivities. This shift in power hits close to home as we consider our pets. It is one thing to accept the subjectivity of a murdered gorilla, a large, powerful, physiologically familiar animal who lives far away; it is another matter entirely to look into our homes at the furry four-legged beasts who have taken up residence there, to reflect on a 30,000-year history of domestication and see not a history of domination but one of mutual constitution.[47]

EATING ANIMALS: MY DELICIOUS PET

While the suggestion that animals are worthy social and historical subjects does not simultaneously require the cessation of the consideration of animals as food, kinship and consumption cannot logically coexist. This contradiction would seem to require resolution. How can we continue to eat animals as we begin to see them as subjects? By simply not eating animals we opt out of the problem that kinship, animal subjects, and cannibalism have presented to us, rather than meeting its uncomfortable challenge. I have suggested that we move into the uneasy realm of the cannibal as a way to seriously explore the way kinship intersects with and conjoins consumption. In this last, brief segment I hope to bring that intersection closer to home.

Besides primates (our genetic kin), domestic animals (specifically pets) are the animals with whom we find it easiest to envision kinship. We trust them and welcome them into our homes. We acknowledge our intimate, shepherding role in their evolution and domestication, and we gloss over our historical motives for desiring domestic animals (easily accessible food, lookouts, and labor) and love them as "furry children."[48] We are thus repulsed by the idea that the Hmong eat cats, that Peruvians eat guinea pigs, that the Dutch eat horses. Despite our studied and objective logic, we are tempted to strip away the humanity, to label as barbaric, those cultures that transgress typical North American bonds of kinship. These present-day Others seem like cannibals to us, ghastly folk who would consume our precious pets. No amount of open-mindedness can convince me that my pet would taste delicious. And yet the arguments above make the contradictory suggestion that in order to truly take this question of animal kin seriously, I must be willing to accept that she might eat me and I her, that we might both be cannibals and that might not be so bad.

Native peoples of North America hunt animals with whom they personally relate and respect.[49] The ontology of these hunting practices mirrors anthropological discussions of reciprocal economies in that the animal is seen to give itself as a gift, and such a gift cannot politely be refused.[50] While not pets, these animals and humans exist in a similarly intimate accord, as part of a network of conscious actors. Animal species understand and accept their fate as the occasional meal in exchange for human respect and gratitude. Carlos Fausto proposes that this "replacement of predation by the gift"[51] is eased by employing a "distinction between consuming the other in its condition as subject and consuming it in its condition as object."[52] But the ability to create an object, ready to be eaten, from a being who was once a subject creates a problem. If kinship can be so easily dismantled when it becomes convenient to look on our kin as prey, then the bonds of kinship are weak indeed.[53] The figure of the cannibal opens up helpful channels from which to consider this problem; the cannibal allows us to see the ways in which consumption of flesh strengthens rather than swallows affinal bonds.

This kind of consumption is transformative and challenging, asking us to accept the consumption of those we consider kin without first adjusting their ontological status: perhaps it is better to consume kin as kin rather than demoting them to objects first. Consumed kin remain kin, are made more kinlike as they are fluidly incorporated into literal and metaphoric acts of cannibalism. Rather than giving in to the gut sense of disgust that the suggestion of cannibalism may invoke, we can embrace the cannibal, a figure who resolves many of the tensions created by seeing animals as subjects, as kin, and as food. Although part of Haraway's definition of a "companion" animal is one who is not eaten,[54] the eaten pet joins the cannibal in my pantheon of discomfiting contradictions that might point the way toward a new, strengthened bond of human-animal kinship. By accepting the poor, beleaguered cannibal, I can begin to confront the suggestion that another creature might be my pet, my kin, and also quite delicious.

NOTES

1. Mark Jenkins, "Who Murdered the Virunga Gorillas?" *National Geographic* 214, no. 1 (2008), 34–65.
2. Jenkins, "Who Murdered the Virunga Gorillas?" 35.
3. Throughout this essay I will be relatively freewheeling with the terms "we" and "our." Rather than qualifying my every sentence, I add here the disclaimer that, as this is a rather personal essay, the "we" I speak of is a particularly North American, academic "we."
4. Jacques Derrida and Elisabeth Roudinesco, "Violence against Animals," in *For What Tomorrow . . . : A Dialogue*, trans. Jeff Fort (Stanford, Calif.: Stanford University Press, 2004), 62–76.
5. Jacques Derrida, "'Eating Well,' or the Calculation of the Subject," in *Points . . . Interviews, 1974–1994* (Stanford, Calif.: Stanford University Press, 1995), 283.
6. Derrida, "'Eating Well,'" 283.
7. Matthew Calarco, *Zoographies: The Question of the Animal from Heidegger to Derrida* (New York: Columbia University Press, 2008), 136.
8. Derrida and Roudinesco, "Violence against Animals," 68.
9. Derrida and Roudinesco, "Violence against Animals," 67.
10. Examples of work exposing the practice of killing and consuming bush-meat include: Dale Peterson, *Eating Apes* (Berkeley: University of California Press, 2003); the photography of Karl Amman; and Donald G. McNeil Jr., "The Great Ape Massacre," *New York Times*, May 9, 1999. Additionally, pages devoted to information about the bush-meat trade exist on the Web sites for the Wildlife Conservation Society (WCS), the Convention on International Trade in Endangered Species of Wildlife (CITES), the Humane Society of the United States (HSUS), and conservation organizations devoted specifically to addressing the bush-meat problem (the Bushmeat Crisis Task Force and the Bushmeat Project).
11. For insight into the sociocultural development of the need for such a mask, see Paula Young Lee, ed., *Meat, Modernity, and the Rise of the Slaughterhouse* (Durham: University of New Hampshire Press, 2008).
12. Lee, *Rise of the Slaughterhouse*.
13. Axel Köhler, "Of Apes and Men: Baka and Bantu Attitudes to Wildlife and the Making of Eco-Goodies and Baddies," *Conservation and Society* 3, no. 2 (2005): 417.
14. Köhler, "Apes and Men," 417.

15. Loretta Cormier, "Monkey as Food, Monkey as Child: Guajá Symbolic Cannibalism," in *Primates Face to Face: The Conservation Implications of Human-Nonhuman Primate Interconnections*, ed. Agustín Fuentes and Linda D. Wolfe (Cambridge: Cambridge University Press, 2002), 79.

16. Maurice Burton and Robert Burton, *International Wildlife Encyclopedia*, vol. 5 (New York: Marshall Cavendish, 2002), 614.

17. John C. Mitani, William J. Sanders, Jeremiah S. Lwanga, and Tammy L. Windfelder, "Predatory Behavior of Crowned Hawk-eagles (*Stephanoaetus coronatus*) in Kibale National Park, Uganda," *Behavioral and Ecological Sociobiology* 49 (2001): 187–195.

18. Vervets, like the rhesus monkeys that are so often the subjects of biomedical research precisely because of their genetic and physiological similarities to us, share 93 percent of their genetic code with humans beings (as compared to 98 percent similarity with the chimpanzee). Nelson Freimer, Ken Dewar, Jay Kaplan, and Lynn Fairbanks, "The Importance of the Vervet (African Green Monkey) as a Biomedical Model," http://www.genome.gov/10002154.

19. The bonobo (*Pan paniscus*) is actually equally close to human beings genetically, but for a number of reasons, both practical and political, they have not received the same kind of popular attention as the chimpanzee (*Pan troglodytes*).

20. David Attenborough, *The Life of Mammals*, BBC Video, DVD, 2003.

21. Donna Haraway, *Primate Visions: Gender, Race, and Nature in the World of Modern Science* (New York: Routledge, 1989).

22. Carlos Fausto, "Feasting on People: Eating Animals and Humans in Amazonia," *Current Anthropology* 48, no. 4 (2007): 4971. See also Matt Cartmill, *A View to a Death in the Morning: Hunting and Nature through History* (Cambridge, Mass.: Harvard University Press, 1996).

23. NOVA's 2009 three-part series *Becoming Human* explores recent scientific research addressing this question. *Becoming Human*, dir. Graham Townsley, 180 min., PBS Video, DVD.

24. Cartmill, *View to a Death*, 226.

25. William Arens, *The Man-Eating Myth: Anthropology and Anthropophagy* (Oxford: Oxford University Press, 1979), 84.

26. Arens, *Man-Eating Myth*, 19.

27. Such labels were, indeed, a mere convenience: Arens concludes that Columbus himself did not actually believe that cannibals inhabited the New World, even as he played his part to propagate these allegations. Arens, *Man-Eating Myth*, 97.

28. Cartmill, *View to a Death*.

29. Arens, *Man-Eating Myth*, 9.

30. Biological anthropologist Phyllis Dolhinow, discussing primate parental care, suggests that "although these patterns of preferential behavior impact fitness, we are not justified in thinking, even for a moment, that it means the animals themselves have a clue as to whether or not an animal is related or whether they themselves or another animal is a parent. How would a nonhuman primate male know he is a father, let alone the father of any specific immature?" While we may recognize familiar patterns of behavior among animals, we must be cautious about ascribing the same sense of motive and understanding to the nonhuman animals who perform these actions. Phyllis Dolhinow, "Anthropology and Primatology," in *Primates Face to Face: The Conservation Implications of Human-Nonhuman Primate Interconnections*, ed. Agustín Fuentes and Linda D. Wolfe (Cambridge, Cambridge University Press, 2002), 19.

31. Fausto, "Feasting," 504.

32. Maggie Kilgour, "The Function of Cannibalism at the Present Time," in *Cannibalism and the Colonial World*, ed. Francis Barker, Peter Hulme, and Margaret Iversen (Cambridge: Cambridge University Press, 1998), 239.

33. Fausto, "Feasting," 508.

34. By a "tradition of reference" I mean that while Arens's denial of the existence of human cannibals (see note 27) has validity, some indigenous cultures still refer to times in their past in which they practiced ritual anthropophagy. While they deny that such a practice persists into the present, the cultural and cosmological beliefs about transformation and affinity that accompany this cultural mythology (be it true or false) are far more interesting and instructive than the debate about what did, did not, or may have happened.

35. Hans Staden, *Han's Staden's True History: An Account of Cannibal Captivity in Brazil* (Durham, N.C.: Duke University Press, 2008), 91.

36. The implication here being that, if eating a human being would be unimaginable, even for a jaguar, then there is something fundamentally *un*natural about the consumption of human flesh.

37. Beth Conklin, *Consuming Grief: Compassionate Cannibalism in an Amazonian Society* (Austin: University of Texas Press, 2001), 187.

38. Staden, *True History*, 52.

39. Staden, *True History*, 59.

40. Staden, *True History*, 131.

41. Cormier, "Monkey as Food."

42. Candace Slater, *Dance of the Dolphin: Transformation and Disenchantment in the Amazonian Imagination* (Chicago: University of Chicago Press, 1994).

43. Beth Conklin, "Hunting the Ancestors: Death and Alliance in Wari' Cannibalism," *Latin American Anthropology Review* 5, no. 2 (1993): 65–70.

44. Conklin, "Hunting the Ancestors," 65.

45. Eduardo Viveiros de Castro, *From the Enemy's Point of View: Humanity and Divinity in Amazonian Society* (Chicago: University of Chicago Press, 1992), 71.

46. Viveiros de Castro, *Enemy's Point of View*, 71.

47. The lighthearted quip, did we domesticate dogs or did they domesticate us? is coy but not vacuous. As animal studies goes about the business of uncovering overlooked animal subjectivities, the prehistorical narrative of animal domestication is no exception. See Tim Ingold, "Making Things, Growing Plants, Raising Animals and Bringing Up Children," in *The Perception of the Environment: Essays in Livelihood, Dwelling, and Skill* (London: Routledge, 2000).

48. Donna Haraway, *The Companion Species Manifesto: Dogs, People, and Significant Otherness* (Chicago: Prickly Paradigm Press, 2003), 11.

49. Conklin quotes Peggy Sanday's comment on Native North American hunting myths: "Just as animals are hunted, so are humans; whoever wants to get food must become food." Quoted in Conklin, "Hunting the Ancestors," 69.

50. Paul Nadasdy, "The Gift in the Animal: The Ontology of Hunting and Human-Animal Sociality," *American Ethnologist* 34, no. 1 (2007): 25–43.

51. Fausto, "Feasting on People," 498.

52. Fausto, "Feasting on People," 513.

53. Erica Fudge provides an instructive discussion of the rabbit. The adorable rabbit, wide-eyed and innocent, disrupts our firm social and linguistic barriers between the pet animal and the food animal. The discomfort at the prospect of consuming such a "cute" animal is enough to compel Fudge's mother to lie to her daughter and call rabbit meat "chicken." While the child openly rejects the idea of eating this particular cuddly animal, the mother's lie reveals a deeper, more adult discomfort that Fudge addresses. Erica Fudge, *Animal* (London: Reaktion, 2002), 34–38.

54. Haraway, *Companion Species*, 14.

PART 2

Applying New Animal Meanings

The Renaissance Transformation of Animal Meaning

From Petrarch to Montaigne

BENJAMIN ARBEL

In his influential book *Animal Liberation* (first edition, 1975), the philosopher Peter Singer raised the idea that the emphasis given in humanistic culture to the dignity of Man, his freedom and his unbound capacities, forcibly led to the decrease of the status of other creatures in Renaissance thought. Singer qualified this generalization by singling out Leonardo da Vinci, Giordano Bruno, and Michel de Montaigne as pioneers in offering an alternative approach to animals.[1] But other scholars who later referred to the historical development of attitudes to animals tended to ignore the evidence of a developing concern for animals in the Renaissance, maintaining instead that Renaissance humanists belittled the value of nonhuman animals.[2]

While the history of attitudes to animals in Europe is now an established area of scholarly inquiry, recent historical writings reflect a near consensus that the Renaissance was an extremely anthropocentric culture that precluded any serious concern for nonhuman animals. According to Erica Fudge, who mainly focuses on late-sixteenth- and early-seventeenth-century England, the paradoxes and contrasts stemming from conceptualizations of the human and the animal in the Aristotelian tradition were the origin of fear and uneasiness rather than of empathy for animals. But the long development of an alternative approach in what the English call "the Continent" is disregarded in her writings.[3] A book dedicated to "Animal and Man in the Thought of the German Renaissance" mainly focuses on religious and theological argumentations concerning the preeminence of Man in God's creation and the beastly part of human nature. Yet it totally ignores the appearance of an alternative current in Renaissance thought about animals.[4] The ideas of Michel de Montaigne, discussed below, are often presented in recent studies as an exception, or alternatively as a beginning of a rethinking about animals,[5] and great emphasis is laid on Descartes' mechanistic theories, which are even occasionally presented as a kind of climax of Renaissance anthropocentrism.[6] It also seems to be widely accepted that more humane attitudes to animals became a significant phenomenon in Europe only from the seventeenth and eighteenth centuries, as expounded, for instance, in Keith Thomas's pioneering book, which also focuses on England.[7] A rare exception in this respect is the chapter dedicated to the Renaissance in Matt Cartmill's book on attitudes to hunting, to which I shall return shortly.[8]

The Renaissance period, especially from the late fifteenth century onward, was characterized in western Europe by demographic and economic expansion, new organization of state institutions (including armies and communication systems), and biological exchange between continents, all of which must have had a negative impact on the fate of animals that came into contact with human society. This chapter, however, does not deal with the treatment of animals in daily life, but rather with changing sensibilities toward animals among Renaissance writers. It is argued here that a new way of considering animals began to develop right from the beginnings of the Italian Renaissance during the age of Petrarch (1304–1374), gaining momentum in the following generations of the Renaissance through the late sixteenth century. The new attitude toward animals was reflected in writings by leading humanists as well as by less famous writers. It was expressed in different ways, such as special treatises devoted to animals, letters, poems, eulogies, epitaphs, and satirical writings, and in passing remarks in works devoted to other issues. But despite their diversity, these writings share some common characteristics, including greater esteem for animals as individuals, a growing appreciation of their mental capacities, and consequently greater attention to moral concern with regard to animals.[9]

The following exposition is not intended to present a comprehensive panorama of all attitudes toward animals that existed or developed in Europe during the Renaissance, but rather to focus on a specific new trend that has hitherto been greatly neglected or totally ignored in historical research. Suffice it to say that the mainstream anthropocentric attitude, which was based on long-rooted traditions and customs, remained largely unchanged, yet as demonstrated below, from the Renaissance onward it had to constantly and continuously defend its human-centered position against a growing number of challengers.

One handicap for understanding the development of Renaissance attitudes to animals stems from the great confusion related to the term "humanism." Considering the rather stereotypical and selective use often made of this term, both in publications treating humanism in general and particularly in works dedicated to Renaissance attitudes to animals, it would be useful to add a brief clarification in this regard. Truly, like most of the "isms" that we use in our languages, "humanism," too, owes its origin to German nineteenth-century intellectuals (as *Humanismus*), and the term "humanism" was widely used by nineteenth- and twentieth-century philosophers. Some of the latter even declared themselves to be or are considered by others as "antihumanists" or "posthumanists." Let me clarify right from the beginning—the later development of the term is of no concern here. This chapter only deals with Renaissance humanism—the important cultural movement that began to develop in the age of Petrarch, reached its high point in Italy during the fifteenth century, and expanded to other European countries during the sixteenth century. Though having a history of its own and consequently being itself subject to change, Italian Renaissance humanism (and to a considerable extent also its continuation in other European areas) had some general characteristics, which have been masterfully analyzed and described by several important scholars, the most prominent of whom is probably Paul Oskar Kristeller. This historian's characterization of Renaissance humanism, which combines clarity of exposition with careful critical qualifications, is briefly expounded in the following paragraphs, particularly for readers unacquainted with the cultural history of the Italian Renaissance.[10]

In the Renaissance period itself the relevant terms employed by contemporaries in relation to what was later described as "humanism" were the Latin terms "*humanitas*," "*studia humanitatis*," or "*humaniores literae*," as well as "*humanista*" (It. *umanista*). In their effort to distance themselves from the traditional scholastic culture, Petrarch and his followers adopted the classical term "*humanitas*" in its Ciceronian significance as "the knowledge of how to live as a cultivated, educated member of society."[11] Their collective effort was based on the rediscovery of the heritage of classical antiquity, particularly those components of this heritage that had been forgotten, neglected, or considerably transformed in the medieval West. Resulting from this cultural movement was a well-defined cycle of studies, the *studia humanitatis* (*humanitas* studies), sometimes also called *humaniores literae*, in the sense of "letters that make you morally better or more human."[12] It comprised grammar (*grammatica*), rhetoric (*rhetorica*), poetics (*poetica*), history (*historia*), and moral philosophy (*philosophia moralis*), which were all studied on the basis of classical texts. These texts were recovered, reedited, translated, and widely distributed, especially after the invention of the printing press around 1450.[13] Consequently, they became part of the basic "cultural baggage" of educated people in Europe for several centuries. Some of these main fields of study, such as history, poetics, and moral philosophy, gained a prominent place in western European culture for the first time since antiquity, whereas the two others—grammar and rhetoric—were considerably transformed compared with their use in traditional medieval culture.

The term "humanists" (Lat. *humanista*, pl. *humanistae*; It. *umanista*, pl. *umanisti*) in the context of Italian Renaissance society and culture denoted persons who had been educated more or less on the basis of this cycle of studies, who were proficient in classical Latin and classical Greek, and who therefore served in several specific capacities such as educators and teachers or as chancellors and secretaries of different states and cities. In other words, this was a kind of profession, especially after the introduction of public lectures in *literae humaniores* in several Italian towns and universities.[14] However, since humanistic education became greatly diffused in the upper and medium echelons of Italian society and later also of other European societies, many men and women, regardless of their occupation or profession, shared the same educational background, including ideas, concepts, terminology, and interests related to *studia humanitatis*. Renaissance humanism was a complex and varied movement that developed over several generations and in different areas. It is not of my concern here to trace this development but only to stress the fact that it is misleading to choose one facet of this movement (even though a major one), such as anthropocentrism, and present it as characterizing the movement as a whole.

As Kristeller emphasizes, the direct contribution of Renaissance humanism to philosophy was concentrated in moral philosophy, an area in which the humanists found themselves in direct conflict and competition with their scholastic contemporaries, since they developed a growing interest in ancient philosophies that were outside the Aristotelian tradition.[15] Moreover, the indirect influence of Renaissance humanism was much wider than the direct one, since humanists made the sources of ancient wisdom and ancient literature available for the first time to their contemporaries, not only in learned editions but also in translations to Latin (in the case of Greek works) and to vernacular languages. These writings, particularly the ones related to moral philosophy, included important works on animal intelligence, on the proper attitude to animals, and even on vegetarianism. The renewed interest in these issues created an intellectual ferment bringing to the fore new ideas and creating new sensitivities, which were also expressed, as we shall see presently, in various literary genres. Interestingly, although central figures in Renaissance humanist culture expressed themselves in one way or another with regard to this topic, none of the leading modern scholars who have studied Italian Renaissance humanism paid any attention to this phenomenon.[16] It is also significant that it took a scholar who had been trained outside this field of studies—a professor of biological anthropology—to discuss this trait of Renaissance culture.[17]

As a matter of fact, humanism was just one component of Renaissance culture, though a very central one. The Renaissance began developing mainly as a literary culture, and its most prominent early protagonists were poets and belles-lettres writers, such as Petrarch, Boccaccio, and others who, though having humanistic education, were not professional humanists. These were arguably the most important promoters of the new sensitivities that are at the center of the present study, which is therefore mainly focused on literary expressions (in the wider sense of this term) of the new attitude to nonhuman beings.

To be able to grasp the changes in this sphere that occurred during the Renaissance it is imperative to consider the ethical consideration of animals in former times. The exploitation of animals for various purposes, such as food, clothing, energy, entertainment, medicine, cosmetics, and not least as scapegoats and sacrifices in religious rites, characterized ancient cultures worldwide. It is therefore no wonder that in both classical cultures and in Judaism, from which Christianity emerged, we can find philosophical and religious concepts aiming at morally justifying such deep-rooted patterns of animal exploitation. Aristotle, for example, expresses a widely accepted division of all creatures into superior and inferior beings, the latter, be they animals or human slaves, having to provide the necessities of the former. Aristotle's treatment of animals in his vast scholarly output is rather complex and sometimes contradictory, but the use that was made (mainly by the Stoics) of his distinction between rational beings, that is humans, and nonrational others, in order to develop a moral code of behavior that only applies to the former, has had an enormous influence on Western culture until this very day.[18]

Yet in the religious and the mythological sphere, as well as in the purely philosophical one, it is also possible to detect opposite trends in ancient cultures. The boundaries between human beings and nonhuman animals and even between gods and nonhuman animals in pagan cultures were not clear-cut; gods were transformed into animals, or existed as composite beings, and the metamorphosis between humans and animals was a central motive in ancient Greek culture. Classical writings portray Pythagoras (ca. 580–490 B.C.E.) and his disciples as believing in the transmigration of souls between animals and human beings, professing vegetarianism, and opposing the abuse of animals.[19] Theophrastus (ca. 372–287 B.C.E.), though an Aristotelian, is reported to have opposed the sacrifice of animals and the consumption of their meat, and even to have contradicted his master's theory that animals were irrational beings.[20] Plutarch (ca. 50–125 C.E.) wrote two treatises in favor of vegetarianism ("On the Eating of Flesh"), another one dealing with the intelligence of animals ("Whether Land or Sea Animals Are Cleverer"), and a parody of the *Odyssey*, entitled "Gryllus," or alternatively "Beasts Are Rational," in which doubts are raised concerning the pretense of Man to be superior to other animals.[21] Porphyry of Tyre (ca. 232–305 C.E.) opposed the killing of animals for food and their use for religious sacrifice. His work *On Abstinence from Killing Animals* is the most comprehensive work written on this subject in antiquity.[22]

These few and nonexhaustive examples are only intended to emphasize that despite the predominant anthropocentrism of ancient Greek and Roman cultures, there was a centuries-long debate on questions concerning the intelligence of animals and, more important, on the ethical obligations of human beings toward them. The absence of any single orthodox religion made such a debate possible. However, the rise of Christianity put an end to such a debate.

Christianity inherited from Judaism a clearly anthropocentric attitude, which can be characterized by the verse in Genesis 9:2–3:

> And the fear of you and the dread of you shall be upon every beast of the earth, and upon every foul of the air, upon all that moveth upon the earth, and upon all the fishes of the sea; into your hand are they delivered. Every moving thing that liveth shall be meat for you, even as the green herb have I given you all things.

Yet whereas in the Old Testament one can find messages that somewhat mitigate the idea of Man's absolute dominance of all other creatures, the New Testament is much less ambiguous in this respect. For example, commenting on the verse from Deuteronomy 25:4 stating: "Thou shalt not muzzle the ox when he tradeth out the corn," Saint Paul comments (Corinthians I, 9:9–10): "Doth God take care for oxen? Or saith He it altogether for our sakes? For our sakes no doubt this is written, that he that ploweth should plow in hope and that he that thresheth in hope should be partaker of his hope." This tendency was reconfirmed and reinforced in the writings of the Church fathers. Though abolishing pagan customs of animal sacrifice and taking distance from other customs such as prophesying by observation of animal intestines, Christianity also put an end to the ongoing debate that took place in the pre-Christian world around the questions of animal intelligence, animal souls, and especially moral concern for animals. This was a gradual process. At the beginning, leading churchmen could still give vent to unconventional views. Saint Basil (ca. 330–379), for instance, could claim that animals did not exist exclusively for the use of Man, having their own raison d'être and their own attachment to "the sweetness of life."[23] It is rather significant that Saint Augustine (354–430) in his *City of God* engaged in polemics with those who claimed that moral rules were applicable to animals. "Not every killing is a murder," he wrote, and the commandment "Thou shall not kill" only applied to human beings. The life and death of all other creatures, according to him, were subject to the necessities of Man by God's commandment.[24] Some centuries later, Saint Thomas Aquinas (1225–1274) wrote in his *Summa Theologiae*, in the spirit of the book of Genesis and Aristotle's *Politics*, that irrational animals (*bruta animalia*) are there to be killed by Man, since the plants and the irrational creatures were created to ensure the existence of humanity. Thus, a person who kills an ox is not guilty of murder, but, at the most, only of causing harm to the property of another person.[25]

In fact, the Aristotelian philosophy, moralized by the Stoics, was particularly fit to sustain Christian dogma in this field, since it considered Man to be the only creature in possession of that part of the soul that contains rational capacities, identified in Christian theology as the very same part of the human soul that remains alive after death. Since animals do not possess it, they have no afterlife and consequently no chance for salvation. The Aristotelian principle, according to which each component of nature has a practical function, was also conveniently used for enhancing this stand, for if animals were destined for human consumption, and since when in paradise human beings are no longer dependent on the necessity to consume food, there would be no use for animals in heaven.[26] Likewise, since it was believed that intelligence and immortality could only occur together, there was little or no use for intelligence among animals.[27]

Thus, the somewhat blurred border between humanity and animality that could often be encountered in classical culture was transformed into an insurmountable wall. Moreover, the ecclesiastical institutions that gradually emerged saw to it that any unorthodox view in this sphere, like in any other, should not find expression in Christian society. It is most remarkable that between the times of Saint Augustine and the early fourteenth century, despite the great importance of animals in medieval culture and literature, hardly any medieval scholar was ready to challenge or question the Christian dogma concerning animals and to defend animals' existence in their own right. Before the mid-fourteenth century it is also rare to find writings or other manifestations expressing emotional feelings toward animals or empathy with animals for their own sake.[28]

Medieval literature abounds with stories about saints who had special relations with animals and could influence them and communicate with them.[29] Yet these cannot be considered as exempla for the proper behavior of ordinary Christians but rather as manifestations of the saintly character of the person concerned, which enabled him or her to act, or rather to perform godly acts, outside the natural, postlapsarian order of things. Recent scholarship has also emphasized the symbolic, allegoric, and metaphoric nature of the hagiographic topoi of saints and animals.[30] The case of Saint Francis is worthy of special consideration. Owing to his preaching to the birds, called by him "my sisters," and his ability to communicate with animals and to influence them, particularly his taming of the wolf of Gubbio, he is nowadays considered to be the patron saint of animals. According to the writings of his disciples, he also tried to convince the civil authorities to ban bird hunting and to provide for the well-being of oxen. These are important messages that have left their mark on posterity. However, as already observed, such actions attributed to this saint were not only considered as miraculous but in this case were also part of his mystical and pantheistic ideas, which also considered the sun, the moon, the wind, and fire as brothers and sisters of Man.[31] Moreover, besides the above-mentioned traditions associated with Saint Francis, there are others that reflect a more anthropocentric attitude, such as the saying attributed by him to every nonhuman creature: "God made me for your sake, O Man."[32] It is also significant that vegetarianism is not included in the Franciscan rule, established by Saint Francis himself for his disciples. No significant school or tradition that developed within the Franciscan order deviated from the orthodox dogma with respect to animals (to date, only one such treatise from the early fifteenth century has come to light).[33] On the other hand, some writings deriving from Franciscan circles include negative expressions concerning animals, such as the reproach expressed by the Franciscan chronicler Salimbene of friars who used to play with cats and dogs, or the decision of the order's council in 1260 prohibiting friars to keep pets in Franciscan monasteries, "except cats and certain birds for the removal of unclean things."[34]

In her book *The Beast Within: Animals in the Middle Ages* Joyce Salisbury has claimed that whereas until the twelfth century the predominant attitude toward animals in medieval culture was one of absolute separation, with all the consequences deriving from this approach in the moral sphere, from the twelfth century onward the walls that had consistently been built by the Church between humans and other animals began to crack. The manifestations of this process were manifold: saints who were presented as free from fear of animals, as protectors of animals, and even as animals, such as the famous Saint Guinefort; the popularity of works presenting the transformation of human beings into animals and hybrid creatures; and the concept of "animality" as something reflecting certain human modes of behavior. In particular, Salisbury attributes great weight to the influence of three greatly diffused literary genres: bestiaries, animal fables, and the popular animal epos *Ysengrimus*, or in its later versions, *Reynard the Fox*, all of which were characterized by anthropomorphism, or the attribution of speech and intelligence to animals and their presentation as a model for human behavior. According to Salisbury, these writings offered an approach different from the orthodox one with regard to animals and contributed to the blurring of the borders between the species.[35]

This shift in the presentation of animals in popular literature could have prepared the terrain for the new trend expressed in Renaissance writings about animals. Yet the fictitious character of fables and similar literary genres in which animals spoke and behaved like human beings was repeatedly emphasized by various medieval writers, such as Isidore of Seville, William of Conches, and John of Garland, and medieval readers were well aware of the fact that bestiaries were didactic rather than scientific writings.[36] Moreover, medieval scholars endeavored to buttress the old Aristotelian division between rational and irrational beings, claiming that animals who were capable of learning only seemed to have something similar to reason, defined by Albert the Great (ca. 1200–1280) as "a shadow of reason." Reason (*ratio*), according to Albert, consisted of two functions: the first involved sense, memory, and *estimativa* (a capacity to elicit intentions); the second involved intellect and was capable of drawing the universal, that is, the principle of art and science. Animals, including pygmies, lacked, according to Albert, this second component. For Saint Thomas Aquinas, animals were completely controlled by their instincts, and any apparent wisdom or prudence in animal behavior was simply a manifestation of God's creative genius.[37]

Very few medieval thinkers were ready to admit that animals had more developed mental capacities. Thus, the twelfth-century scientist Adelard of Bath contended that animals had the power of judgment (*iudicium*) since they altered their behavior according to their interpretation of outside signals.[38] Roger Bacon (ca. 1214–1294) admitted that animals possessed a mode of cognition entailing a process resembling logical discourse (*quasi quoddam genus arguendi*) and that they could recognize universals and draw conclusions, but he also insisted that only humans were truly capable of knowledge and reason.[39] At any rate, both of them did not draw any moral conclusions from such reflections, which were formulated rather carefully so as not to transgress Christian dogma. The only serious challenge to medieval orthodoxy in this sphere, as in several others, was that of the Cathari, who believed in transmigration of souls between human beings and animals. But the Cathari were condemned as heretics and were brutally suppressed in the thirteenth century.[40]

Of course, animals occupy an important role in the literature of hunting, which appeared in Europe in growing numbers from the thirteenth century. A considerable part of that kind of writing was dedicated to hunting dogs, and some of these texts are virtual panegyrics for these animals. The most famous one is probably the hunting book written by Gaston Phebus, Count of Foix, toward the end of the fourteenth century.[41] Sophia Menache has claimed that in this genre, one can follow a shift from an instrumental attitude to animals to a genuine expression of affection toward them.[42] This seems to be true, though I would suggest that rather than a new attitude we are here confronted by a new capacity or readiness to express certain feelings in writing. It is also to be remembered that hunting and killing animals—which, after all, was what that genre was all about—remained not only legitimate but an important status symbol of the aristocratic milieu for which this kind of literature was written.

The main tenets concerning the treatment of animals, as defined in the early stages of Christianity and buttressed during the long medieval period, predominated also in early modern European culture both in the Catholic and the Protestant camps. The deep-rooted human customs based on the exploitation and killing of animals continued to be supported by a complex system of myths, beliefs, philosophies, and laws that were meant to justify these customs. The symbolic, allegorical, and didactic traditions related to animals that had developed during the Middle Ages also had a central role in many manifestations of Renaissance culture, particularly in Renaissance art.[43] This notwithstanding, during the Renaissance there appeared a growing number of literary works expressing empathy for animals as such, disregarding the ecclesiastical dogma by recognizing animals' mental capacities, and claiming that animals were entitled to moral consideration. It is also significant that such arguments were generally not expounded in religious terms.

At the root of this change was the development of Renaissance humanism, characterized by a growing interest in ethics and a marked secular trend, based on the rediscovery of classical cultures, particularly of those parts of ancient written heritage that had been forgotten or neglected in the Latin West for many centuries. The rediscovered classical writing that influenced Renaissance rethinking about animals included, for example, Plutarch's tractate on animal intelligence, his two compositions on vegetarianism, and his dialogue "Gryllus" (to which we shall return shortly); several of Plato's dialogues; works by Theophrastus, who claimed that animals resembled human beings in their emotions, perceptions, and reasoning, and were therefore worthy of justice; Porphyry's work *On Abstinence from Animal Food*; Lucretius's *On the Nature of Things*; the philosophical traditions attributed to Pythagoras (particularly their presentation in Ovid's *Metamorphoses*, 15.75–142); and the skeptic writings of Sextus Empiricus. Another manifestation of this growing interest in animals was the enormous popularity in Renaissance culture of Pliny's *Natural History*, a work that disregarded the philosophical traditions that refused to recognize the mental capacities of animals.[44]

The artistic manifestations of the phenomenon treated here are too numerous, too complex, and too interesting to be contained in a few paragraphs. But it is important to remember that Renaissance art is simply flooded with animals of all sorts. Truly, many of these have a symbolic and/or allegorical function, as do their medieval predecessors, but in many other cases, and sometimes even in the very same ones, they reflect a new sensitivity and new readiness to express bonds of affection between human beings and other animals. In Renaissance art animals not only function as symbols and allegories but also appear as real living beings alongside their human companions and sometimes even independently.

In the literary sphere, as in many other manifestations of Renaissance culture, we have to start with Petrarch. The poet's strong attraction to nature, predating the romantics of the eighteenth and nineteenth centuries, is quite well known. Less known is his predilection for dogs, and especially the literary expression of this characteristic trait. In a letter written in 1347 to his patron, Cardinal Colonna, he described his life with a big white dog of Spanish origin presented to him by the cardinal. Here is one stanza of this enchanting testimony:

> . . . I live in freedom now,
> Since he and only he alone is my protector,
> And my companion constantly. At night,
> When wearied with the labors of the day,
> I seek my couch and close my eyes in sleep,
> He guards my house; and if I sleep too long
> He whimpers, telling me the sun has risen,
> And scratches at my door. When I go forth
> He greets me joyously, and runs ahead
> Toward places often visited, and turns
> Around from time to time to see if I
> Am following; and then, when I recline
> Upon the verdant margin of the stream
> And there begin again my wonted task,
> He starts this way and that, tries all the paths,
> And then lies down, white on the grassy ground,
> Turning his back to me, his face to those
> Who may pass by . . .[45]

Petrarch's letter carries us one step beyond the dog literature of the aristocratic hunting culture. It expresses a harmonious relationship between dog and master, independent of any social or symbolic connotations. Keeping animals as companions and having emotional attachment to them is a universal phenomenon, and was certainly not new in the age of Petrarch.[46] What was new is the cultural milieu that enabled Petrarch and his successors to give literary expressions to such feelings.

From the times of Petrarch onward, classical models of writing poems, eulogies, and epitaphs dedicated to animals inspired Renaissance writers to do the same. Those very few examples of animal epitaphs or eulogies composed during the Middle Ages are most probably of symbolic significance.[47] During the Renaissance, first in Italy and then in other European countries, the classical tradition of empathic writing about animals was revived among writers of both sexes who wrote panegyrics, epitaphs, and eulogies dedicated to their companion animals.[48] Petrarch wrote a short epitaph for his little dog Zabot (not the protagonist of the above-cited letter).[49] Isabella d'Este, marchioness of Mantua, engaged a group of poets in 1511 to compose epitaphs in memory of Aura, her beloved little bitch.[50] Quite famous are also the three poetic epitaphs written by the French poet Joachim du Bellay (1522–1560) to a dog, to his cat, and to another small dog.[51] Some modern scholars excluded the possibility of a serious motivation behind these writings, describing them as a rhetorical exercise, as a "mock encomium," a "mock oration," a "comic dialogue," or as manifestations of a literary fashion that had no serious intent behind it.[52] It is true that some Renaissance scholars clearly exaggerated in describing the extraordinary talents of their pets. Thus, Leon Battista Alberti (1404–1472), one of Italy's most famous and creative humanists, wrote a eulogy in memory of his dead dog in which he claimed that his companion animal was not only a sagacious and moral being but that he was also proficient in the liberal arts.[53] The humanist Jacob Locher from Nuremberg (1471–1528) dedicated the last pages of his book (1506) to his beloved bitch, who, according to him, authored the Latin verses in which she presented herself to the reader.[54] Such humorous references are similar to those we use nowadays to express admiration for the intelligence of our beloved companion animals. It is to be assumed that contemporary readers who were amused by reading such writings were also able to distinguish between truth and rhetorical embellishment.

Even if in some cases such writings were indeed written without any serious intent, one cannot escape the impression that many of them express candid friendship and intimate affection in a way that could not even be imagined in medieval culture. Moreover, even as rhetorical exercises, some of them reveal acquaintance with classical philosophies that cannot be reconciled with Christian dogma (such as those of Pythagoras and Porphyry), sensitivity to the suffering of animals, appreciation of their individual personalities, and remorse for maltreatment of animals, as expressed, for instance, in the eulogy for an ass composed by Laura Cereta (1469–99), a young woman from Brescia.[55] There is also no reason to doubt the serious intent behind the decision of János Zsámboky (known as Johannes Sambucus, 1531–1584), the Hungarian court physician of Emperor Maximilian II, to include in his successful book of emblems one dedicated to his two dogs, Madel and Bombo. In the accompanying text he wrote: "These dogs merit their fame. Why, then, would you deny that these companion animals have reason inside?" (*sensus in esse cur neges his bestiis sequacibus?*)[56] The author's portrait included in the second edition of this work also includes Bombo, whose name appears under this dog's figure (see figure 1). The interest in eulogies and monuments for companion animals was apparently quite diffused in the early sixteenth century, considering that later editions of the popular lexicon *Cornucopiae epitome*, published under the name of the French humanist Jean Textier, Seigneur de Ravisy (known as Ravisius Textor, 1480–1524), also included lists of famous dogs and horses.[57]

We can also detect interesting insights reflecting a reconsideration or reevaluation of the nature and virtues of animals compared with those of human beings in the unfinished poetic

Figure 1. Johannes Sambucus (János Zsámboky) with his dog, Bombo. Johannes Sambucus, *Emblemata*, 2nd ed. (Antwerp: Plantin, 1566). (Source: Leiden University Library, Thysia, 1197, 2. By permission)

fable "The Ass," written by Niccolò Machiavelli (1469–1527). Machiavelli used a theme developed in Plutarch's "Gryllus," the above-mentioned parody of Homer's *Odyssey* in which a person transformed by the sorceress Circe into a pig expresses the reluctance of Odysseus' companions to return to human guise, and when offered the opportunity to do so enumerates the misery of the human condition compared with that of other animals. In Machiavelli's poem, the pig (actually a wild boar) is expressing the writer's pessimistic view of the human character, which can also be found in his political writings. The poem has been interpreted as a refutation of that idealistic trend in humanistic culture that exalted what was considered to be the unique supernatural capabilities of human beings. It also expresses a relativistic world-view criticizing human anthropocentrism.[58] What makes this worldview so pessimistic is the idea that human beings are doomed to suffer forever from their unstable character, from their enslavement to their ambitions and uncontrolled desires. According to the boar's speech, all other animals have natural virtues and are free from those human weaknesses. No wonder, therefore, that the beastly interlocutor in Circe's menagerie did not have any intention of returning to his former human status, as expressed in the last stanzas of the poem:[59]

> No pig causes pain to other pigs,
> No deer to other deer; only Man
> Kills, crucifies and robs other men.
> . . . And if any among human beings seems to you divine,
> Happy and delighted, do not believe him too much,
> For I live much happier in this mud,
> In which I bathe and roll without worries.

One of Machiavelli's private letters also reveals his sensitivity with respect to animals. In his response to a letter from his son Guido, in which Guido complained about the behavior of a young mule kept in the family farm outside Florence, Machiavelli gave the following advice to his son:[60]

> As for the little mule, who seems as if he had run mad, it is advisable to treat him in a manner different from that which is usually applied to madmen. The latter are normally tied up, and I would like you to let him free. Let Vangelo conduct him to Montepulciano, remove his rein and halter, and allow him to go wherever it pleases him, to eat by himself and get rid of his madness. The land is big, the animal small and unable to harm anybody. Thus, without bridle, it will be possible to observe what he wants to do, and once recovered you may take him back.

Machiavelli's consideration of the mule's psychological problem and the way he proposes to handle it is more than just a pretty anecdote in the biography of the famous political thinker. He recognizes the individual personality of the mule as well as his idiosyncratic preferences and expresses his dissatisfaction with the customary methods of punishment and violence as treatment of stubborn animals, and probably also of insane people.

A few decades later one can find a similar attitude in the Italian literature on horse training. The first author who seems to have taken distance from the customary violent methods in this field was Claudio Corte, whose book was first published in Venice in 1562. In England, John Astley criticized violent methods of horse breaking in his book published in 1582.[61] It is even possible that the new pedagogical theories developed by Italian humanists eventually left their mark also on the taming of beasts.[62]

The new sensitivities among Renaissance writers brought up another important theme—criticism of hunting. Hunting was a central component in medieval and Renaissance culture, and during the Renaissance it arguably reached unprecedented dimensions in royal courts and aristocratic milieus. Criticism and even prohibition of hunting during the Middle Ages can be found in ecclesiastical documents. It mainly concerned the hunting activities of churchmen but also, though more mildly, those of laymen. However, the arguments raised in these medieval sources were theological and were not related to the moral consideration of animals.[63] As already emphasized by Matt Cartmill, Renaissance criticism of hunting was a new phenomenon, different from its medieval predecessors.[64] Renaissance humanists and other writers developed a secular criticism of hunting based on moral arguments that expressed an aversion to this occupation simply because it entailed cruelty to animals.

Sebastian Brant's critical remarks in his *Ship of Fools* (*Narrenschiff*, 1494) can still be considered as a continuation of the traditional criticism of hunting, presenting as a model of behavior Saint Hubert and Saint Eustace, both of whom abandoned hunting in order to serve God (paradoxically, these two figures are now considered to be patron saints of hunters).[65] But during the very same time humanists who passed criticism on their societies included hunting among the targets of their disparagement and did it without using any religious argumentation. Thus, in chapter 39 of his *Praise of Folly* (1508), Erasmus ironically refers to the attachment of aristocrats to hunting:[66]

> This class of madness also includes those who look down on everything except hunting wild animals and whose spirits are incredibly exhilarated whenever they hear the nerve-shattering blasts on the horns or the baying of the hounds. I imagine that even the dung of the dogs smells like cinnamon to them. And then what exquisite pleasure they feel when the quarry is to be butchered! Lowly peasants may butcher bulls and rams, but only a nobleman may cut up wild animals. Baring his head and kneeling down, he takes a special blade set aside for that purpose (for it would hardly do to use just any knife) and exercises the most devout precision in cutting up just these parts, with just these movements, in just this order. Meanwhile, the surrounding crowd stands in silent wonder, as if they were seeing some new religious ceremony, although they have beheld the same spectacle a thousand times before. Then, whoever gets the chance to taste some of the beast is quite convinced that he has gained no small share of added nobility. Thus, though these men have accomplished nothing more by constantly chasing and eating wild animals than to lower themselves almost to the level of the animals they hunt, still in the meantime they think they are living like kings.

The citizens of Thomas More's *Utopia* (first edition, 1516) considered hunting to be an occupation unworthy of free people and left it to their butchers who were slaves. Here is how More describes the Utopians' feelings about hunting:[67]

> What pleasure can there be in listening to the barking and howling of dogs—isn't that rather a disgusting noise? Is any more pleasure felt when a dog chases a hare than when a dog chases a dog? If what you like is fast running, there's plenty of that in both cases; they're just about the same. But if what you really want is slaughter, if you want to see a creature torn to parts under your eyes—you ought to feel nothing but pity when you see the little hare fleeing from the hound, the weak creature tormented by the stronger, the fearful and timid beast brutalised by the savage one, the harmless hare killed by the cruel hound. And so the Utopians, who regard this whole activity of hunting as unworthy of free men, have accordingly assigned it to their butchers, who, as I said before, are all slaves. In their eyes, hunting is the lowest thing even butchers can do. In the slaughterhouse their work is more useful and honest, since there they kill animals only out of necessity; whereas the hunter seeks nothing but his own pleasure from killing and mutilating some poor little creature. Taking such relish in the sight of slaughter, even if only of beasts, springs, in their opinion, from a cruel disposition, or else finally produces cruelty, through the constant practice of such brutal pleasures.

One can detect in this text an influence of the traditional Thomistic attitude (repeated acts of cruelty to animals brutalize human character),[68] but also a clear aversion to the killing of "harmless" creatures by "cruel" ones with human encouragement. In one of More's Latin poems a hunter "looks and smiles" as his hound tears a rabbit to pieces. "Insensate breed," he writes, "more savage than any beast, to find cruel amusement in bitter slaughter!"[69] We also know that this English humanist kept a small menagerie in his home, including a little monkey who is depicted beside Dame Alice, More's wife, in Holbein's sketch prepared for More's family portrait (1527).[70]

Michel de Montaigne (1533–1592; see figure 2), the French aristocrat, admits in his essay "On Cruelty" that he also took part in hunting expeditions. Yet at the same time he writes:[71]

As for me, I have not even been able to witness without displeasure an innocent defenseless beast which has done us no harm being hunted to the kill. And when as commonly happens the stag, realizing that it has exhausted its breath and its strength, can find no other remedy but to surrender to us who are hunting it, throwing itself on our mercy which it implores with its tears:

> *All covered with blood, groaning, and seeming to beg for grace*
> [Virgil, *Aeneid*, VII, 501].

That has always seemed to me the most disagreeable of sights.

Then he adds:

I hardly ever catch a beast alive without restoring it to its fields. Pythagoras used to do much the same, buying their catches from anglers and fowlers.

It was, I think, by the slaughter of beasts in the wild that our iron swords were first spattered with warm blood [Ovid, *Metamorphoses*, XV, 106–107].

Figure 2. Portrait of Michel de Montaigne (ca. 1590). (Source: Private collection. Courtesy of Philippe Desan)

Montaigne concludes this section, writing: "Natures given to bloodshed where beasts are concerned bear witness to an inborn propensity to cruelty."

A similar censure of aristocratic engagement in hunting can be found in Miguel de Cervantes's *Don Quixote* (first edition, 1605). Don Quixote and his squire, Sancho Panza, were invited to take part in an aristocratic hunting party. The event did not prove successful for Sancho, who, when climbing a tree in his escape from a wild boar, unluckily tumbled and ended up swinging in midair with his head downward. No wonder that the criticism of hunting is put by Cervantes in Sancho's mouth: "Why should kings and other great folks run themselves into harm's way, when they may have sport enough without it: Mercy on me! What pleasure can you find, any of ye all, in killing a poor beast that never meant any harm!"[72] When the duke insists on explaining the virtues of hunting Sancho remains unconvinced, concluding that hunting "goes mightily against my calling and conscience." Cervantes also wrote the charming novel *The Dialogue of Dogs*, in which sixteenth- and early-seventeenth-century Spanish society is depicted as perceived through the eyes of two dogs. Human cruelty to animals is one of the themes treated in this novel.[73]

The concept of cruelty was rarely treated in medieval literature. Thomas Aquinas does dedicate a chapter to it in his *Summa*, where it is of course interpreted in Christian terms as something that concerns the soul, as a result of a conscious and rational act on the part of human beings. However, in another important work Aquinas explains that biblical passages forbidding cruelty to "dumb animals" do not reflect consideration for animals but are rather aimed at preventing the brutalizing of human character, so as to preclude cruelty to other human beings, and also at preventing temporal loss for the animal's owner.[74] Montaigne is the earliest modern writer who dedicated a special essay to cruelty. Yet he does not relate to it in theological terms, and for him cruelty is embodied in the action itself.[75] Moreover, the examples he uses do not merely refer to relations between humans but also to how people behave to animals: ripping off a hen's head, watching hounds tearing a hare to pieces, or enjoying the killing of a stag by hunting dogs.

While art is not my focus in this chapter, I cannot resist mentioning in this context a work of art by Albrecht Dürer (1471–1528) (who worked in close collaboration with humanists) of a dead deer's head pierced by a hunter's arrow (see figure 3). It cannot be excluded that this was the artist's protest against the hunt. According to Colin Eisler, who has investigated Dürer's animal paintings: "Although much of Dürer's work is devoted to suffering, to Christian torments in often repetitive detail, few of these come close to the sense of loss and shocked witness provided by the stricken deer. It is as if he wanted to come to graphic terms with wanton destruction. A whole world seems to have died with this deer."[76]

Let us now return to the issue of relativism in Renaissance culture. We have already encountered this approach in Machiavelli's "The Ass" and in Cervantes' *Dialogue of Dogs*, both of which challenged the prevailing anthropocentric culture. In his *Praise of Folly* Erasmus of Rotterdam includes similar reflections when he ironically refers to the classical (and also modern) claim that animals were inferior beings because they lacked the ability to speak or were ignorant of grammar. Judging by what Erasmus put in Folly's mouth, the horse who does not master grammar or does not eat cakes has no reason to be miserable since Man also cannot fly, and even a stupid person is not miserable since this is his or her nature. Moreover, "the grammar of even one language is more than enough to make life a perpetual agony."[77] Thus the very characteristic singled out by philosophers as an exclusive capacity of human beings is here presented as a source of misery. The motive of happiness and misery is repeated in another chapter, where Erasmus puts the following phrases in Folly's mouth (or pen):[78]

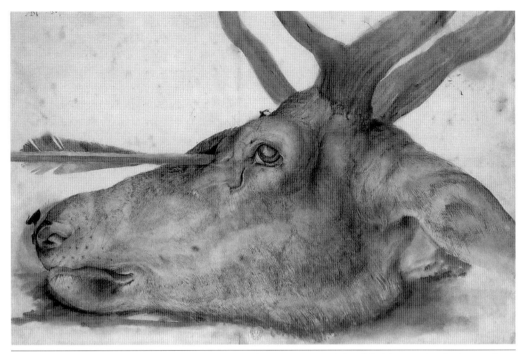

Figure 3. Albrecht Dürer: Deer's head pierced by an arrow. (Source: Courtesy of the Bibliothèque nationale de France)

> I could never bestow sufficient praise on that cock embodying Pythagoras, who had been, in his single person, a philosopher, a man, a woman, a king, a private citizen, a fish, a horse, a frog, even a sponge (I think), but who decided that no creature was more miserable than Man because all the others were content to remain within the limits of Nature, while Man alone tried to go beyond the bounds of his lot.

These lines reflect Erasmus's acquaintance with the Pythagorean tradition, but also, like Machiavelli's "The Ass," are reminiscent of Plutarch's "Gryllus." The repudiation of idealistic humanism combined with a new approach to the concept of happiness (expressed in secular terms), in which animals are also included, seems to have appealed to sixteenth-century writers and readers.

Letting Folly express such ideas should not mislead us into taking them lightly. In fact, parody was becoming an increasingly popular genre to convey unorthodox ideas in the age of the Counter-Reformation. This way was also chosen by Giambattista Gelli, the humanist cobbler of Florence, in his work entitled *Circe*, first published in Italian in 1549. Like the above-quoted passages from Machiavelli's "The Ass," this work, too, is a parody of the *Odyssey*, or rather another elaboration of Plutarch's parody, "Gryllus." Gelli developed the ancient parody by further elaborating the plot. In *Circe*, each one of Odysseus's companions was transformed by the sorceress into a different animal. When confronted by Odysseus's proposal to return to human form, each animal expounded the advantages of its proper animal state in comparison with the disadvantages of being human. However, Gelli was apparently less pessimistic than Machiavelli, since in his dialogue the elephant, originally a philosopher, finally agreed to return to his human status.[79]

Gelli's dialogue enjoyed wide distribution. Besides five sixteenth-century editions in

Italian (some in several printings), it was published in English in 1557 (two more editions followed in the eighteenth century), French (1569), Latin (Vienna 1595), and German (1620). Most probably, many readers considered it to be a mere literary entertainment, but some at least must have read it in the light of the relativist philosophy that gained momentum among sixteenth-century writers. One of the exponents of this new trend was even a bishop, Gerolamo Rorario, who served as nuncio of Pope Clement VII in the Hungarian court. His work, entitled *Animals Often Use Reason Better Than Man*, was written in Latin in 1544.[80] Rorario's relativism is even more radical than the above-mentioned parodies, since he dwells on the issue of animal intelligence, challenging the traditional division between rational and irrational beings. No wonder that the papal nuncio never dared to make his thoughts public. His book was printed posthumously in 1648 and brought considerable fame to its author, and Pierre Bayle even dedicated to him a long article in his *Dictionnaire historique et critique* (1696–1697).[81] In the meantime, owing to Descartes and his disciples, the questions concerning animal thinking, animal souls, and the moral consequences deriving therefrom had reached the forefront of intellectual debate.

As a matter of fact, the relativist trend had reached its climax already in the sixteenth century in the *Essays* of Michel de Montaigne, first published in 1580. In the essay "On Cruelty" Montaigne wrote that after learning about the resemblance of animals to human beings and the many common traits they share, one had to lower considerably the human pretence of dominion over other creatures.[82] But most of his relativist ideas concerning the human-animal divide appear in his longest essay, "An Apology for Raymond Sebond," where Montaigne draws a great number of examples from classical literature, especially from Pliny the elder and Plutarch, concerning the abilities of animals. Yet unlike the authors of medieval bestiaries, which still enjoyed some popularity during his times, he never tries to use this material allegorically or analogically for didactic purposes. What he seeks in these sources are proofs and examples showing that abilities that were considered to be exclusively human were in fact shared by other animals. These abilities include intelligence and rational thinking, imagination, mutual communication, memory, feelings, emotions (such as love and loyalty), the ability to learn, the ability to dream, and even the ability to exercise belief. Human pretense to be the sole creature on Earth enjoying such faculties is presented by Montaigne as vainglory. As far as he is concerned, all creatures, including human beings, act out of instincts and are endowed with different measures of intelligence.[83] Referring to the question of language, which was and still is a central issue in the debate about animal intelligence, he writes (following Plutarch): "We can only guess whose fault it is that we cannot understand each other: for we do not understand them any more than they understand us. They may reckon us to be brute beasts for the same reason that we reckon them to be so."[84] Montaigne is convinced that animals can communicate not only with members of their own species but also with other animals. He draws our attention to the many forms of human communication that are not part of written or spoken language and are very similar to the way animals communicate with one another. He does not deny that humans enjoy certain mental capacities that set them apart from other animals, yet for him these are not the most important qualities of life. The French humanist's relativism is in a way summarized in the following famous sentences: "When I play with my cat, how do I know that she is not passing time with me rather than I with her? We entertain ourselves with mutual monkey tricks. If I have times when I want to begin or to say no, so does she."[85]

Though being active in the same century, Niccolò Machiavelli and Michel de Montaigne did not belong to the same generation. They lived in different countries, and their social backgrounds were also quite different. Yet both of them were products of humanistic education. They were both very well versed in the classical sources and had what can be described as an intimate relationship with classical authors. Both of them were central and well-known figures of late Renaissance culture. But what is most relevant to our subject is their pessimistic outlook with regard to human nature, which in both cases may have resulted from the respective political background of their adult life: the Italian wars in Machiavelli's case and the French wars of religion in Montaigne's. And what is even more remarkable is that both Machiavelli (in a minor way) and Montaigne (quite blatantly) drew similar conclusions with regard to animals from their pessimistic humanism. The words of the boar in Machiavelli's poetical fable "The Ass" point to a relativistic attitude of the Florentine intellectual, and his letter concerning his mule reflects recognition of animals' individual personality and consideration for their needs. In the case of Montaigne we pass to radical and even subversive attitudes that are expounded much more extensively and also include the recognition of animals as moral subjects.

This change in sensitivities concerning animals in Renaissance culture is also reflected in Renaissance travel literature, particularly in writings about the Islamic East. Whereas the moral questions concerning a more "humane" attitude to animals hardly appears in medieval travelogues, the greater consciousness in Renaissance culture to moral questions related to the proper treatment of animals brought several travelogue writers to pay attention to this issue when observing and describing the mores and customs that they encountered in the East. They were generally impressed by the favorable treatment of animals in those regions, and their descriptions were later used by other writers as examples for the proper way of treating animals in Europe itself.[86]

In conclusion, as in many other spheres, beside phenomena reflecting continuity and conservatism in attitudes to nonhuman beings, the Renaissance period also represents a significant new shift toward open empathy for animals and awareness of human responsibility for them in modes, contexts, and dimensions that had not existed in Europe since antiquity. In various ways and forms, writers active between the age of Petrarch and that of Montaigne expressed a new sensitivity regarding animals, recognized their mental and emotional capacities, and also acknowledged a moral obligation toward them. It is to be assumed that during the Middle Ages as well, people of various classes, especially those who lived in close contact with animals, experienced similar feelings about and impressions from animal behavior. But only during the Renaissance did a cultural milieu develop that enabled educated people, particularly laymen with humanistic education, to give vent to such thoughts and feelings in writing. They did so by using secular arguments, distancing themselves from Christian orthodoxy concerning the essential differences between animals and human beings. The presentation of Renaissance humanism as a current that contributed to the degradation of the condition of animals in the West is therefore unjustified.

The criticism against Man's cruelty to animals often came from the same scholars who also criticized other aspects of their societies, such as war, slavery, and the attitude to other human races. It was precisely this small but not marginal group of humanists who opened a public debate concerning the intelligence of animals, putting in question the unbounded exploitation of other species. This debate could take place during the Renaissance because lay society gained sufficient strength and self-confidence to voice unconventional ideas, whether directly or in the form of parodies and paradoxes. The Renaissance writers that have been the subject of this chapter inaugurated a public debate that continues today.

So, did animals have a Renaissance? If we mean the animals themselves—the wolves, foxes, bears, and other wild animals who continued to be exterminated throughout Europe, or the pigs, calves, hens, fish, and other living creatures that continued to be systematically killed for daily human consumption—the answer is evidently no. The exceptions were those relatively few animals who were lucky enough to be treated like Petrarch's dogs, Machiavelli's mule, or Montaigne's cat. But if we mean by this question what important Renaissance intellectuals thought and wrote about animals, then the answer is yes. Inspired by the rediscovery of long-neglected classical writings, particularly those related to animals; taking a fresh look at other writings that have always been there; and using such materials to take distance from traditional attitudes in a more secular manner, some of them adopted a rather skeptical attitude to long-established anthropocentric concepts. The change was there, but history does not evolve linearly, and it would take a long time until these changing sensitivities would be translated into organized action, and even then—or better, even now—we are still very far from a significant change in the traditional anthropocentric patterns of thought and behavior that characterize Western attitudes to animals.

NOTES

1. Peter Singer, *Animal Liberation*, 2nd ed. (New York: New York Review, 1990), 198–200.
2. This common image is repeatedly expressed in references to the Renaissance included in books dedicated to animal rights or animal liberation, mostly written by authors with a philosophical background; see, for instance, Mary Midgley, *Animals and Why They Matter* (Athens: University of Georgia Press, 1983), 11; Richard D. Ryder, *Animal Revolution: Changing Attitudes towards Speciesism* (Oxford: Basil Blackwell, 1989), 43–45; Jim Mason, *An Unnatural Order: Why We Are Destroying the Planet and Each Other* (New York: Continuum, 1993), 34–35.
3. Erica Fudge, *Perceiving Animals: Humans and Beasts in Early Modern English Culture* (London: Macmillan, 2000); Erica Fudge, *Brutal Reasoning: Animals, Rationality, and Humanity in Early Modern England* (Ithaca, N.Y.: Cornell University Press, 2006).
4. Maria Suutala, *Tier und Mensch im Denken der deutschen Renaissance* (Helsinki: SHS, 1990). I am grateful to Tom Tyler for turning my attention to this book.
5. Thierry Gontier, *De l'homme à l'animal: Montaigne et Descartes ou les paradoxes de la philosophie moderne sur la nature des animaux* (Paris: Librairie philosophique J. Vrin, 1998); Brian Cummings, "Animal Language in Renaissance Thought," in *Renaissance Beasts: Of Animals, Humans, and Other Wonderful Creatures*, ed. Erica Fudge (Urbana: University of Illinois Press, 2004), 179–182; Stefano Perfetti, "Philosophers and Animals in the Renaissance," in *A Cultural History of Animals in the Renaissance*, ed. Bruce Boehrer (New York: Berg, 2007), 163–164.
6. James Serpell, *In the Company of Animals: A Study of Human-Animal Relationships* (Oxford: Basil Blackwell, 1986), 124.
7. Keith Thomas, *Man and the Natural World: Changing Attitudes in England, 1500–1800* (London: Penguin Books, 1983). See also Peter Harrison, "The Virtues of Animals in Seventeenth-Century Thought," *Journal of the History of Ideas* 59 (1998): 463–484.
8. Matt Cartmill, *A View to a Death in the Morning: Hunting and Nature through History* (Cambridge, Mass.: Harvard University Press, 1993), 76–91.
9. The artistic expressions of this trend, as well as its reflection in works of natural history, will be treated separately in future essays.

10. Kristeller's historization of humanism has been described by another important scholar as "probably the most important revolution in post-war scholarship on Renaissance thought." See James Hankins, "Two Twentieth-Century Interpreters of Renaissance Humanism: Eugenio Garin and Paul Oskar Kristeller," in *Humanism and Platonism in the Italian Renaissance*, vol. 1, *Humanism* (Rome: Edizioni di Storia e Letteratura, 2003), 588. It is significant, but also rather upsetting, that a recent and apparently successful introduction to humanism does not even mention the works of Kristeller or those of other important historians of Italian humanism, such as Eugenio Garin and Richard Trinkaus, to mention just a few. See Tony Davies, *Humanism*, 2nd ed. (New York: Routledge, 2008). Finally, the peculiar way in which the term "humanism" is used in modern Italian historiography to denote a distinct period (predating the Renaissance) will not be adopted here.

11. John F. D'Amico, *Renaissance Humanism in Papal Rome* (Baltimore: Johns Hopkins University Press, 1983), 124.

12. Hankins, "Two Twentieth-Century Interpreters," 574.

13. For a succinct presentation of the characteristics of these fields and the innovative approach to them in Renaissance culture, see Paul Oskar Kristeller, "Humanism," in *The Cambridge History of Renaissance Philosophy*, ed. Charles B. Schmitt and Quentin Skinner (Cambridge: Cambridge University Press, 1988), 113–137.

14. These terms served as semantic roots upon which the later term "*humanismus*" was built.

15. Kristeller, "Humanism," 134–135; Paul Grendler, *Schooling in Renaissance Italy: Literacy and Learning, 1300–1600* (Baltimore: Johns Hopkins University Press, 1989), 11–141.

16. For example, Eugenio Garin, *Italian Humanism: Philosophy and Civic Life in the Renaissance* (Westport, Conn.: Greenwood Press, 1965); Paul Oskar Kristeller, *Renaissance Thought and Its Sources* (New York: Columbia University Press, 1979); Richard Trinkaus, *The Scope of Renaissance Humanism* (Ann Arbor: University of Michigan Press, 1983); Anthony Grafton and Lisa Jardine, *From Humanism to the Humanities* (Cambridge, Mass.: Harvard University Press, 1986). In the index of Schmitt and Skinner's *Cambridge History of Renaissance Philosophy*, the entry "animals" is only followed by a cross-reference to "zoology." Likewise, most of the material treated in a recent essay on attitudes to animals in the Renaissance concerns writings on natural history, besides a short survey of translations of ancient zoological writings, as well as a page and a half on Montaigne's "Theriophily"; see Perfetti, "Philosophers and Animals in the Renaissance," 147–164.

17. Cartmill, *View to a Death*, 76–91.

18. For an excellent introduction to moral attitudes to animals in ancient Greek and ancient Roman philosophy, see Richard Sorabji, *Animal Minds and Human Morals: The Origin of the Western Debate* (Ithaca, N.Y.: Cornell University Press, 1993). For a shorter critical presentation, see Stephen T. Newmyer, "Animals in Ancient Philosophy: Conceptions and Misconceptions," in *A Cultural History of Animals in Antiquity*, ed. Linda Kalof (Oxford: Berg, 2007), 151–174.

19. Jo-Ann Shelton, "Beastly Spectacles in the Ancient Mediterranean World," in Kalof, *Cultural History of Animals in Antiquity*, 111–112; Newmyer, "Animals in Ancient Philosophy," 151–152, 155–156.

20. Newmyer, "Animals in Ancient Philosophy," 163.

21. Plutarch, "Whether Land or Sea Animals Are Cleverer," "Beasts Are Rational," and "On the Eating of Flesh, I–II," in *Moralia*, vol. 12, ed. G. P. Goold (Cambridge, Mass.: Harvard University Press, 1995), 311–479, 489–533, 537–579. See also Newmyer, "Animals in Ancient Philosophy," 169–172.

22. Porphyry, *On Abstinence from Killing Animals*, trans. Gillian Clark (London: Duckworth, 2000); see also Newmyer, "Animals in Ancient Philosophy," 172–173.

23. Ryder, *Animal Revolution*, 34.

24. Saint Augustine, *The City of God*, ed. Philip Scharff, in *A Select Library of the Christian Church: Nicene and Post-Nicene Fathers*, vol. 2 (Peabody, Mass.: Hendrickson Publishers, 1995), 14–15.

25. Saint Thomas Aquinas, "Homicide," in *Summa Theologiae: Latin Text and English Translation, Introductions, Notes, Appendices and Glossaries* (New York: McGraw-Hill, 1975), 38:19–21.

26. Joyce Salisbury, *The Beast Within: Animals in the Middle Ages* (New York: Routledge, 1994), 76.

27. Peter G. Sobol, "The Shadow of Reason: Explanations of Intelligent Animal Behavior in the Thirteenth Century," in *The Medieval World of Nature: A Book of Essays*, ed. Joyce E. Salisbury (New York: Garland, 1993), 121–122.

28. One such evidence is the poetical eulogy for a dead dog written by Theodoric of St. Trond (near Liège). See Frederic J. E. Raby, *A History of Secular Latin Poetry in the Middle Ages*, 2nd ed., vol. 2 (Oxford: Clarendon Press, 1967), 144–145.

29. See, in particular, Helen Waddell, ed. and trans., *Beasts and Saints*. Woodcuts by Robert Gibbings (London: Constable, 1934); Dominic Alexander, *Saints and Animals in the Middle Ages* (Woodbridge: The Boydell Press, 2008).

30. Brigitte Resl, "Introduction: Animals in Culture, ca. 1000–ca. 1400," in *A Cultural History of Animals in the Middle Ages*, ed. Brigitte Resl (New York: Berg, 2007), 17; Alexander, *Saints and Animals*.

31. Ernesto Grillo, ed., *Early Italian Literature*, vol. 1, *Pre-Dante Poetical Schools* (London: Blackie, 1920), 145–146; Edward Armstrong, *Saint Francis, Nature Mystic* (Berkeley: University of California Press, 1979); Roger D. Sorrel, *Saint Francis of Assisi and Nature: Tradition and Innovation in Western Christian Attitudes to the Environment* (New York: Oxford University Press, 1988); Alexander, *Saints and Animals*, 169–180.

32. Singer, *Animal Liberation*, 197–198.

33. Ryder, *Animal Revolution*, 33.

34. Serpell, *In the Company of Animals*, 48.

35. Salisbury, *Beast Within*; on Saint Guinefort, see Jean-Claude Schmitt, *The Holy Greyhound: Guinefort, Healer of Children since the Thirteenth Century*, trans. Martin Thom (Cambridge: Cambridge University Press, 1983).

36. Jan M. Ziolkowski, *Talking Animals: Medieval Latin Beast Poetry, 750–1150* (Philadelphia: University of Pennsylvania Press, 1993), 16–19, 23.

37. Albertus Magnus, *On Animals: A Medieval Summa Zoologica*, trans. and ed. Kenneth F. Kitchell Jr. and Irven Michael Resnick (Baltimore: Johns Hopkins University Press, 1999), vol. 2, book 21, 1416–1417. See also Sobol, "Shadow of Reason," 109–128.

38. Adelard of Bath, *Conversations with His Nephew: On the Same and the Different, Questions on Natural Science and On Birds*, ed. and trans. Charles Burnett (Cambridge: Cambridge University Press, 1998), 111–119; Francis Klingender, *Animals in Art and Thought to the End of the Middle Ages*, ed. Evelyn Antal and John Harthan (Cambridge: MIT Press, 1971), 347–348.

39. Sobol, "Shadow of Reason," 116.

40. Malcolm Barber, *The Cathars: Dualist Heretics in Languedoc in the High Middle Ages* (Harlow, U.K.: Longman, 2000).

41. Gaston Phebus, *The Hunting Book of Gaston Phebus: Manuscrit Français 616*, introduction by Marcel Thomas and Francois Avril, ed. Wilhelm Schlag (Paris: Bibliothèque Nationale, 1998).

42. Sophia Menache, "Hunting and Attachment to Dogs in the Pre-Modern Period," in *Companion Animals and Us: Exploring the Relationships between People and Pets*, ed. Anthony L. Podbersek, Elizabeth S. Paul, and James A. Serpell (Cambridge: Cambridge University Press, 2000), 42–60.

43. Simona Cohen, *Animals as Disguised Symbols in Renaissance Art* (Leiden and Boston: Brill, 2009).

44. For the views of these classical writers on animals, see Sorabji, *Animal Minds and Human Morals*. On the popularity of Pliny's *Natural History* in the Renaissance, see Brian Cummings, "Pliny's

Literate Elephant and the Idea of Animal Language in Renaissance Thought," in *Renaissance Beasts*, ed. Erica Fudge (Chicago: University of Illinois Press, 2004), 171; Eugene Willis Gudger, "Pliny's *Historia Naturalis*: the Most Popular Natural History Ever Published," *Isis* 6, no. 3 (1924): 269-281.

45. Petrarch, *Selected Sonnets, Odes and Letters*, ed. Thomas G. Bergin (New York: Appleton-Century-Crofts, 1966), 127–130 (*Ep. Met.* III, 5).

46. For the keeping of companion animals during the Middle Ages, see Serpell, *In the Company of Animals*, 47–49.

47. See, for instance, Ziolkowski's discussion of Sedulius Scottus's "The Ram" (mid-ninth century), in his *Talking Animals*, 69–79.

48. For example, Jacob Burckhardt, *The Civilization of the Renaissance in Italy*, trans. S. G. C. Middlemore (1860; repr., London: Phaidon Press, 1951), 158; René de Maulde la Clavière, *The Women of the Renaissance* (London: J. Allen, 1900), 407; Jan Papy, "Lipsius and His Dogs: Humanist Traditions, Iconography and Rubens's Four Philosophers," *Journal of the Warburg and Courtault Institutes* 62 (1999): 167–182; Cristiano Spila, ed., *Cani di pietra: L'epicedio canino nella poesia del Rinascimento*, trans. by Maria Gabriella Critelli and Cristiano Spila (Rome: Quiritta, 2002). I am thankful to Paolo Procaccioli for bringing the last publication to my attention.

49. Morris Bishop, *Petrarch and His World* (London: Chatto & Windus, 1964), 289.

50. Spila, *Cani di pietra*.

51. Joachim du Bellay, *Poésies françaises et latines*, ed. E. Courber (Paris: Librairie Garnier Frères, 1931), 305–306, 356–361, 352–355.

52. Such explanations were already given in the seventeenth century as a defense for Justus Lipsius's (1547–1606) epitaph for his dog; see Papy, "Lipsius and His Dogs," 170. For some recent examples of such an interpretation, see Albert Rabil Jr., *Laura Cereta, Quattrocento Humanist* (Binghamton, N.Y.: Center for Medieval and Early Renaissance Studies, 1981), 52 ("a mock funeral oration"); Diana Rubin, ed. and trans., *The Collected Letters of a Renaissance Feminist* (Chicago: University of Chicago Press, 1997), 16 ("a comic dialogue"). See also Gontier's refutation of similar claims with regard to Montaigne, in Gontier, *De l'homme à l'animal*, 42, 86, 88, 95.

53. Cecil Grayson, ed. and trans., "Il *Canis* di Leon Battista Alberti," in *Miscellanea di studi in onore di Vittore Branca, Umanesimo e Rimascimento a Firenze e Venezia* (Florence: Leo Olschki, 1983), 193–204; Leon Battista Alberti, *Il Cane*, trans. [in the late fifteenth century] from Latin into Italian by Piero di Marco Parenti Fiorentino (Ancona: Tipografia di Aurelj G. e comp., 1847).

54. Colin Eisler, *Dürer's Animals* (Washington, D.C.: Smithsonian Institution Press, 1991), 168.

55. Rabil, *Laura Cereta*, 118–134, 202.

56. Johannes Sambucus, *Emblemata* (Antwerp: C. Plantin, 1564), 164–165. See also A. S. Q. Visser, *Joannes Sambucus and the Learned Image: The Use of the Emblem in Late-Renaissance Humanism* (Leiden: Brill, 2005); Antonio Bernat Vistarini, Emilio Blanco, John T. Cull, and Tomás Sajó, "His Master's Voice: Johannes Sambucus and His Dog Bombo," *Silva. Digital Review of Studiolum*, 3, December 15, 2004, http://www.studiolum.com/en/silva3.htm.

57. Bernat Visantini et al., "His Master's Voice," 1. The lists appear in a posthumous edition that is also attributed to Texier.

58. Paolo Fazion, "L'asino da leggere," in *Machiavelli: L'Asino e le bestie*, ed. Paolo Fazion and Gian Mario Anselmi (Bologna: Cooperativa Libraria Universitaria Editrice, 1984), 25–134.

59. Niccolò Machiavelli, "Dell'Asino d'oro," in *Tutte le opere*, ed. Francesco Flora and Carlo Cordiè, vol. 2 (Milan: Arnoldo Mondadori Editore, 1960), 781 (my translation).

60. Niccolò Machiavelli, *Lettere*, ed. Franco Gaetà (Milan: Feltrinelli, 1961), No. 231 (my translation).

61. Frederick Smith, *The Early History of Veterinary Literature and Its British Development*, vol. 1, *From the Earliest Period to a.d. 1700* (London: J. A. Allen, 1976) 138, 148, 190; Joan Thirsk,

"Horses in Early Modern England: For Service, for Pleasure, for Power," in *The Rural Economy of England: Collected Essays*, ed. Joan Thirsk (London: Hambledon Press, 1984), 389–390.

62. Keith Tomas attributed such an influence to a later period; see his *Man and the Natural World*, 45.

63. Thomas Szabó, "Die Kritik der Jagd, von der Antike zum Mittelalter," in *Jagd und höfischer Kultur im Mittelalter*, ed. Werner Rösener (Göttingen: Vandenhoek & Ruprecht, 1997), 170–175, 177–230; Harald Wolter-von dem Knesebeck, "Aspekte der höfischen Jagd und ihrer Kritik in Bildzeugnissen des Hochmittelalters," in Rösener, *Jagd und höfischer Kultur im Mittelalter*, 523–529.

64. Cartmill, *View to a Death*, 76–91.

65. Eisler, *Dürer's Animals*, 97.

66. Desiderius Erasmus, *The Praise of Folly*, ed. and trans. Clarence H. Miller (New Haven, Conn.: Yale University Press, 1979), 60–61.

67. Thomas More, *Utopia*, trans. Robert M. Adams, ed. George M. Logan and Robert M. Adams, rev. ed. (Cambridge: Cambridge University Press, 2002), Book II, 70–71.

68. Serpell, *In the Company of Animals*, 153 (citing Aquinas's *Summa contra gentiles*, III, 112). See also text with note 74.

69. More, *Utopia*, 71, note 75 (quoted by the editors).

70. Richard Marius, *Thomas More: A Biography* (Cambridge, Mass.: Harvard University Press, 1999), 229–230. The finished painting was later destroyed by fire.

71. Michel de Montaigne, *The Complete Essays*, trans. M. A. Screech (London: Penguin, 1991), 484–485 [book 2, essay 11].

72. Miguel de Cervantes, *Don Quixote*, Ozell's revision of the translation of Peter Mottcux (New York: Modern Library, 1950), 673–674.

73. Miguel de Cervantes, "The Dialogue of Dogs," trans. Lesley Lipson, in *Exemplary Stories* (Oxford: Oxford University Press, 1998), 250–305.

74. Serpell, *In the Company of Animals*, 153 (citing Aquinas's *Summa contra gentiles*, III, 112).

75. Daniel Baraz, "Seneca, Ethics and the Body: The Treatment of Cruelty in Medieval Thought," *Journal of the History of Ideas* 59, no. 2 (1998): 195–215.

76. Eisler, *Dürer's Animals*, plate 11 and text on 102. See also Cartmill, *View to a Death*, 80–81.

77. Erasmus, *Praise of Folly*, 51.

78. Erasmus, *Praise of Folly*, 53.

79. Giambattista Gelli, *La Circe* (Florence: Torrentino, 1549). For an English translation, see *The Circe of Signior Giovanni Battista Gelli etc.*, trans. Thomas Brown, with an introduction by Robert Adams (Ithaca, N.Y.: Cornell University Press, 1963).

80. Hieronymus Rorarius, *Quod animalia bruta saepe ratione utantur melius homine*, ed. Gabriel Naudé (Paris: Cramoisy, 1648); Luigi Perissinotto, "Perché gli animali spesso usano la ragione meglio dell'uomo," *Rivista di estetica* 8, no. 38 (1998): 177–196; Aidée Scala, *Girolamo Rorario: Un umanista dipolomatico e i suoi dialoghi* (Florence: Olschki, 2004), 115–162.

81. Pierre Bayle, *Historical and Critical Dictionary: Selections*, ed. and trans. Richard H. Popkin (New York: Bobbs-Merrill, 1965), 214–254.

82. Montaigne, *Complete Essays*, 487.

83. Montaigne, *Complete Essays*, 489–683, esp. 505–541. For an excellent analysis of this important essay, see Gontier, *De l'homme à l'animal*, 39–158.

84. Montaigne, *Complete Essays*, 506.

85. Montaigne, *Complete Essays*, 505.

86. Benjamin Arbel, "The Attitude of Muslims to Animals: Renaissance Perceptions and Beyond," in *Animals and People in the Ottoman Empire*, ed. Suraiya Faroqhi (Istanbul: Eren, 2010), 57–74.

On the Trail of the Devil Cat

Hunting for the Jaguar in the
United States and Mexico

SHARON WILCOX ADAMS

Few animals capture the human imagination like large mammalian predators. Charismatic megafauna such as bears, wolves, and big cats are frequently present in cultural discourse, working as symbols and proxies for a vast range of beliefs, values, taboos, and anxieties. The largest feline species in the Western Hemisphere, the jaguar (*Panthera onca*) has long evoked complex and often conflicting emotions of awe, reverence, anger, and fear in neighboring human communities.[1] In areas now identified as the southwestern United States and Mexico, the jaguar was represented as a symbol of hunting prowess, spirituality, divine might, political power, and/or military strength among different cultural groups for thousands of years.[2] Through visual art, folklore, mythology, rituals, and the act of killing the physical animal, humans have sought to understand, represent, and possess the jaguar.

Building from the body of scholarship establishing the prominence and significance of the jaguar in pre-Columbian cultures, this chapter is concerned with accounts of jaguars produced by naturalists and hunters in the nineteenth and early twentieth centuries.[3] Examining how the jaguar was encountered, identified, and subsequently represented in these narratives reveals not only the bodies of physical animals but also a symbol intricately connected to the ways in which humans understand and communicate about wildlife and animality. By recording observations in field notes and travel logs, telling jaguar-lore around the campfire, and tracking the elusive feline through remote regions, hunters and naturalists traversed landscapes of jaguar-occupied spaces both real and imagined.

When looking at writings from men working in this region, the categories of hunter, travel guide, scientist, or naturalist are not exclusive. Indeed, the professional and personal identities of a number of these individuals span several, if not all, of these categories. For the purpose of this chapter, the individuals discussed in the following sections are identified by the activities they were engaged in at the time of their jaguar encounters. American and European men largely produced the sources examined in this study. Certainly, this racial and gendered homogeneity in the production of these sources must be carefully interrogated, as there are a number of silences in these documents drawn along lines as diverse as species, nationality, gender, class, and race. Reading through these accounts and narratives offers insight into the ways the jaguar is imbued with notions of social class, prestige, whiteness, and masculinity.

WRITING ALONGSIDE THE JAGUAR

Writing a social history centered on an animal, particularly a rare wild animal, presents a number of challenges.[4] Animals themselves leave little record of their existence, save for the occasional skeletal remains. Natural processes in the environment typically erase other identifiable traces, such as the remains of prey kills, scat (feces), and tracks (paw prints). In order to uncover animals in the past, we must turn to places where we may uncover what geographer Michael Woods eloquently terms "ghostly representations" that "speak on behalf" of the animal, such as scientific reports, written accounts, folklore, oral histories, religion, and art.[5]

Representations of animals, in historical documents or present-day discourse, should not be confused or conflated with the animal subject itself. Human knowledge of animals is not informed through access or insight into the perspective of the "real" animal, or the ways in which an animal might perceive and live its own reality.[6] Instead, humans are limited in the ways in which they are able to understand and communicate about the *idea* of the animal, *as* humans, and within individual, overlapping social discursive networks. Representation, or re-presentation, is a subjective and dynamic process, whereby certain markers are abstracted from the body or life of the "real" animal in order to identify and communicate about the species. Seemingly objective sources—including scientific reports, photographs, and videos—work alongside more apparently subjective sources—such as folklore, narrative film, art, and literature—to contribute to the construction of what Bruno Latour terms "immutable mobiles" that provide members of society with standardized representations of the animal subject.[7] These "immutable mobiles" enable participation in the ongoing social construction of the animal within a society or social group, even among those who have no direct experience with an animal.[8] These representations are "immutable" or unchanging in meaning, while also possessing a certain fluidity as they are contextualized within the time and place they are produced, reproduced, deployed, or interpreted.

While representation is a necessary and vital part of human communication, it must be considered critically. This process cannot be characterized as simple reproduction; rather, Woods argues, it is a process of translation whereby an animal subject is detached from the representation, taking on a new, entirely different form as an object.[9] These representational objects are inherently contestable, as this process of translation divorces the subject from the ability to exert any control over its own representation and places the power of construction and deployment within the human communities who utilize these representations as a way to communicate about the subject.[10] Scholarship within disciplines examining the history of science offers a vast range of examples demonstrating that forms of representation that are commonly thought to be objective or unbiased, such as scientific reports, photographs, and video, must be understood to be very much human enterprises and therefore subjected to perspectives, motivations, and the deeply situated knowledge of those who produce these representations—be they transparent or opaque, deliberate or subconscious.[11]

Through the processes of representation, jaguars are removed from their own animality, or jaguar-selves, and enter cultural discourse as objects.[12] This process must be understood not as a reproduction of the jaguar-as-animal, but as a translation whereby the jaguar assumes a new form imbued with human notions of jaguar-ness. The meanings and values coded within this spotted body have changed many times over thousands of years, negotiated and renegotiated relative to social, economic, and political contexts. These representations are extremely powerful, significantly shaping human actions and policies toward the "real" animal on the landscape.

Before examining representations of the jaguar in human-produced reports, drawings, and narratives, this chapter first foregrounds the animal itself through a brief discussion of the jaguar's biology, ecology, and geographic range. While the enterprise of scientific description itself is a representational practice that must necessarily abstract the animal, it is fitting to start here in order to provide a useful method for framing experience and expectation of encounter with physical animal bodies.

NATURAL HISTORY OF THE JAGUAR

The jaguar is the largest cat species native to the Americas and the third largest in the world.[13] Jaguars are commonly identified by their distinctive spotted coat, which is often confused with that of the leopard (*Panthera pardus*), a species native to the Eastern Hemisphere.[14] While both cats have a similar brownish/yellow base fur color with dark rosette markings, the jaguar can be distinguished by the presence of small dots or irregular shapes within the larger rosette markings, a more stocky and muscular body with a larger and broader head, relatively shorter limbs, larger paws, and a shorter tail.[15] The jaguar measures five to eight feet from nose to tail and weighs 140 to over 300 pounds, with females slightly smaller than males.[16]

While jaguars are believed to prefer areas with good vegetative cover near rivers, streams, or other wetlands such as dense forests or swamps, they can be found in a wide range of habitats, ranging from tropical forests to semidesert grasslands.[17] In Sonora, Arizona, and New Mexico, the mountainous terrain of Madrean evergreen woodland has historically supported jaguars, although this population has been significantly less dense here than in the lush forests in the southern part of Mexico and farther down the species range into Brazil.[18]

Home range varies widely among the species, from 25 to 40 square kilometers in size for females to roughly double that area for males.[19] The abundance of prey species, the availability of space, and the presence of other jaguars all affect the size of these ranges.[20] Typically, male jaguars' home ranges do not overlap with those of other males, except in areas of abundant prey.[21] The prey base of jaguars is extensive and opportunistic, as jaguars take full advantage of the diversity of animal species found throughout their variety of habitats. Overall, more than eighty-five species have been recorded in the jaguar's diet, including peccaries (wild pigs), capybaras (large rodents), deer, sloths, caymans, tapirs, monkeys, freshwater fish, birds, reptiles, amphibians, and other small animals.[22] Jaguars will also hunt larger animals, including domestic stock if readily available, which historically fostered significant trouble with ranchers throughout their range and resulted in representations of the cat as a villainous predator and thief who needed to be "brought to justice."[23]

GEOGRAPHIC DISTRIBUTION

The historical range of the jaguar extends from the central regions of South America, throughout Central America, and north into the United States.[24] The jaguar species *Panthera onca* is recorded throughout Pleistocene fossil records, first appearing approximately 1.8 million years B.P. and spanning the North American continent from as far north as present-day Washington,

Nebraska, and Pennsylvania and as far east as Florida.[25] While fossil records locate the jaguar in Florida 7,000 to 8,000 years ago, an account from 1711 places "tygers . . . differ[ent] from the Tyger of Asia and Africa" in the mountains part of the Carolina colony.[26] As discussed later in this chapter, this conflation of "Old World" nomenclature with "New World" fauna has lead to debate as to whether these "tigers" were jaguars, mountain lions, or even bobcats. While the jaguar's range retracted southward throughout the nineteenth century, reported sightings still placed jaguars in Arizona, New Mexico, Texas, and possibly Southern California, Colorado, and Louisiana.[27]

By the mid-nineteenth century, the greatest threats to jaguar populations were (and continue to be today) human encroachment and resulting development and deforestation, which drastically affected the jaguar's prey base and fragmented the cat's population into isolated pockets, rendering them vulnerable to a number of threats and pressures.[28] Hunting, both for sport and by ranchers and private landowners who viewed the jaguar as a nuisance species, was also a significant threat to jaguar populations.[29] The rate of jaguar sightings grew increasingly rare, and David E. Brown and Carlos A. Lopez Gonzalez identify a total of sixty-four jaguars recorded in Arizona and New Mexico between 1848 and 2001.[30] While the methods with which sightings are confirmed and counted may be debated or contested, these records do reveal that the jaguars on the northernmost edge of the population were increasingly vulnerable to threats and pressures, leading to a contraction in range south of the political border. Thus, those seeking any sort of jaguar experience were compelled to journey south from the United States into Mexico.

WHAT'S IN A NAME?

One of the most significant challenges to resurrecting jaguars from narratives and scientific reports of the past is the common confusion of terminologies used to identify cat species in the American Southwest and Mexico. Further complicating this confusion in nomenclature is the difficulty many experience in correctly identifying an animal "in the field." The species *Panthera onca*, or what is today known as the "jaguar," has been identified by a wide range of terms over the past 500 years. The term "jaguar" is commonly reported to have its origins in the Tupi-Guarani Indian name "*yaguara*," which has been translated a number of ways, including "eater of us," "body of a dog," and the most commonly cited, "the beast (or animal) that kills its prey in one leap (or bound)."[31] However, Brown and Lopez Gonzalez note that this more "exotic" name of "jaguar" has only recently come into use in the southwestern United States in the mid-twentieth century.[32]

In the nineteenth and early twentieth centuries, the jaguar was also commonly referred to as the "American leopard" or "Mexican leopard," or, more commonly, by the term still used in Mexico today, "*el tigre*" (tiger).[33] By utilizing these names for species from Asia, Europe, and Africa, early Europeans absorbed these new animals into their existing worldview. Not only did this have the effect of confusing historical records through inconsistent use of terminology, but there is also a significant power dynamic in appropriating "New World" animals with "Old World" labels (this legacy proved to complicate species identifications through the nineteenth century).[34] Through the acts of identification and classification, what was new became familiar and what was unknown became "known." This process of colonial reimagining negates symbolic meanings constructed by American Indians, enforcing a degree of hegemonic control over production of Eurocentric conceptions of nature through the body of the jaguar.

Steve Pavlik establishes the myriad ways in which indigenous feline terminology, nomenclature, iconography, and folklore have been translated in a manner that, without proper cultural contextualization, can be confusing and even misleading.[35] Pavlik considers the work of Karl Luckert, a scholar of Navajo religion who attributed the use of the term "tiger" in Navajo hunter mythology as a reference to the Asian tiger, suggesting that the animal was added to narratives of origin when its existence became known to the Navajo.[36] However, Pavlik argues, "it seems much more probable that the Navajos adopted the word for tiger from the Spanish and Mexicans with whom they had contact for more than three hundred years and who used *tigre* in reference to the jaguar."[37] Both Pavlik and Nicholas Saunders assert that the prevalence of the term "tiger" among Native cultures throughout the region is not the result of remarkable similarities in cosmology between groups, but rather a postcontact development reflecting complicated interactions between Europeans and American Indians, whereby European interpretations of feline imagery may have been oversimplified and essentialized. In particular, European translation seemed largely ignorant of nuanced differences between real world and spiritual animal beings, calling attention to the crucial role of cultural context in understanding representations.[38] Compellingly, Saunders argues "conquest" may have elicited "revitalized and accentuated" feline imagery within American Indian cultures, perhaps as a way to embody "jaguar-ness" as a form of spiritual resistance to "conquest" as well as a reaction to subsequent postcontact events such as widespread smallpox pandemic.[39]

In Brazil, the jaguar is known as *onça*.[40] This term, used by Europeans in South America, formed the origins of the jaguar's Latin name, *Panthera onca*. However, the term "*onza*" has a very complex history in Mexico, where it is used to identify both the snow leopard (*Uncia uncia* or *Panthera uncia*) and as a general term for medium-size cat species.[41] Ernesto Alvarado Reyes notes that the term "*onza*," when combined with the common name of a species, is a way in Mexico to indicate a variety of species with "recessive traits that make them look different to most individuals from their population."[42] This colloquialism is likely the reason that A. Starker Leopold documented this term being used in rural northern Mexico to describe jaguarundi (*Felis yagouaroundi*).[43] This terminology is still further complicated by the fact that in northwestern Mexico, the term "*onza*" refers to a mythological large wildcat that inhabits the Sierra Madre Occidental. Neil Carmony describes this fabled animal as "not a jaguar, not a mountain lion, the onza was considered more elusive and ferocious than either."[44] Certainly, the wide range of contexts within which this term was deployed lead to great confusion in communicating about cats' presences on the landscape and in cultural discourse.

Adding further confusion to the identification of species is the reliability of eyewitness accounts. There are a number of cat species whose range overlaps to some degree with the jaguar, and some are far more common. These include the mountain lion (*Felis concolor*), which is commonly referred to as the "*leon*" (lion) in Mexico, while in the United States the animal is commonly known as "mountain lion," "cougar," "panther," "catamount," or "puma."[45] Similar in size to the jaguar, the mountain lion is nonetheless easy to distinguish, as adults do not possess a spotted coat and while the juveniles are spotted, these markings occur in three irregular dorsal lines rather than the distinctive rosettes of the jaguar.[46] More commonly smaller spotted cat species including ocelots (*Felis pardalis*) and margays (*Leopardus wiedii*)are confused for juvenile jaguars.[47] Additionally, the terms *tigrillo* ("little tiger") and *oncillo* ("little onca") are also used to describe these smaller spotted cats.[48] The terms "*pantera*," "*leopardo*," and "*onza*" are used throughout writings about Mexican wildlife, often used interchangeably to describe any and all of these species.[49]

Further complicating this problematic nomenclature are melanistic (black) jaguars, which are often confusingly referred to as "*pantera*" in Mexico, or "black panthers," a name that is also applied to black leopards and black mountain lions. Despite a rash of unconfirmed sightings in the borderlands, black jaguars are not known to occur north of Belize.[50]

Encounters with wildlife are typically fleeting (unless an animal is killed as a result), also making it difficult to accurately identify an animal in the wild. Brown and Lopez Gonzalez note, "That most people want to see a jaguar greatly increases the incidence of misidentification and normally reliable people have made jaguars out of large dogs (especially yellow or black Labrador retrievers) and even house cats and coatis."[51] As the next section demonstrates, identification was not only difficult for those working amongst feline species in the field; it was also a challenge for those attempting to standardize these conventions in the research facilities of Europe.

NATURALIST NARRATIVES

Since first contact with the New World, European scholars were fascinated with the project of creating systems to organize and classify a flood of new discoveries and observations.[52] One of the most widely respected resources for the classification of animals was French naturalist and zoologist Georges Cuvier's *Regne animal distribué d'après son organisation* (*The Animal Kingdom Arranged in Conformity with Its Organization*), a series of four volumes published in 1817 (a second edition of five volumes was published in 1829–1830).[53] Cuvier relied on descriptions and specimens collected from the field, since he preferred not to travel far from his offices at the Jardin des Plantes in Paris, France. Cuvier drew from his career's work of researching the structure of living and fossil animals to develop his reference series, considered the authority in zoology until Charles Darwin published *The Origin of Species* in 1859. Including written descriptions and detailed illustrations of all known species, each subsequent edition included newly discovered and reclassified organisms in keeping with the rapid changes occurring in zoology throughout the nineteenth century. Within these volumes, Cuvier's work reveals a lively and ongoing discourse regarding the origin, classification, and distribution of species encapsulated within these representations of feline-ness.

The volume concerned with mammalian species includes two illustrations of spotted large cats identified as "*The Jaguar—Felis onca*," accurately illustrated with the cats' stocky confirmation and coat pattern of open rosettes (the name "jaguar" had already gained popularity in Europe at this time to describe the species).[54] The entry includes a brief description of the species' spotted coat, "a lively fawn colour above; the flank with four rows of ocellated spots that is with rings more or having a black point in the middle white beneath with black," while also commenting on the species perceived threat to humans, stating "nearly the size of the Royal Tiger and almost as dangerous."[55] The geographic range for the species is not given.

A second set of illustrations of large spotted cats also corresponds with this section, labeled "*The Panther—Felix Pardus–the Pardala of the Ancients.*" The entry describes a cat "fawn coloured above white beneath with six or seven rows of black spots resembling roses that is formed by the assemblage of five or six simple spots on each flank the tail is the length of the body minus that of the head."[56] This entry also notes that the species is found in Africa, Asia, and India. This cat has a physical build longer and leaner than the jaguar and a spotted coat. This "Panther of the Ancients" appears to be a modern-day leopard. However, this is complicated by the following entry, "*The Leopard—Felis leopardus*" and the brief description of the cat being from Africa and "similar to the Panther but has ten rows of smaller spots." Cuvier includes in his notes, "these two species (Leopard and Panther) are smaller than the Jaguar. Travelers and furriers designate them indiscriminately by the names leopard, panthers, African tiger, etc."[57]

Cuvier's entries reveal an ongoing debate that spanned over 100 years as to "[w]hether the Leopard and the Panther are in reality a distinct species, and if so, on what particular characters the specific distinction depends" as well as the place of the jaguar within this order.[58] Cuvier's classification of panthers was an attempt to address observable variations in the species known today as "leopard." His work builds from and in response to earlier classifications by both Carolus Linnaeus, in his seminal *Systema Naturae*, and Georges-Louis Leclerc de Buffon, a French naturalist whose thirty-five-volume *Histoire Naturelle* was published 1749–1788. According to Cuvier, Buffon's identification of the spotted cat species went awry, causing great confusion. Buffon placed "jaguars" in the Eastern Hemisphere (it is believed Buffon confused the jaguar with the species Cuvier called "Panther of the Ancients"). The British *Penny Cyclopedia* comments on this matter:

> It may not however be useless to observe, that of the figures given by Buffon as Panthers and Jaguars, that which is entitled the male Panther is in all probability a Leopard; the female is unquestionably a Jaguar; the Jaguars of the original work and of the supplement, are either Ocelots or Chatis; and that which purports to be the Jaguar or Leopard, although probably intended to be a Chetah, is not clearly referable by its form and markings to any known species. . . . When therefore it is said that the Panther much resembles the Jaguar, it is always to be strongly suspected that the type whence the observations have been taken is in reality an American animal.[59]

Clearly, even when the animal specimens, be they live or dead, are at one's disposal, it does not necessarily help with this larger-scale confusion of species. Buffon's errors would inspire nearly 100 years of debate, as other naturalists such as Cuvier sought to apply their own corrections to these errors through their own publications.

Cuvier's volumes, notes, and the ongoing discourse from other works reveal the enthusiasm and confusion with which New and Old World species were identified and connected. The text in Cuvier's volume spoke little to the lives of these species, as the specimens were observed in zoos, menageries, and offices and not in their natural environment. Perhaps the most lasting contribution of *The Animal Kingdom* are the lithograph illustrations, as the colorful images set the imagination ablaze. Cuvier represents the animal bodies standing still or as though moving at a relaxed gait, but never frozen in a moment of predatory impulse. The volume includes two illustrations of jaguars. One jaguar is depicted with his mouth open. His teeth are not bared; rather, his tongue protrudes and is reminiscent of the "flehmen face" where a feline curls its lip in order to facilitate the transfer of pheromones and other scents into the vomeronasal organ, also called the Jacobson's organ (see figure 1).[60] The second jaguar also has what seems to be an uncannily human expression on his face (see figure 2).

Similarly, the two illustrations of the "Panther of the Ancients" are evocative. The "Panther of the Ancients" stands with a quiet nobility and unusually expressive eyes, while the second illustration shows a panther with an almost anthropomorphized expression of mild humor as it stares directly at the viewer (see figures 3 and 4). Cuvier's work did much to demythologize

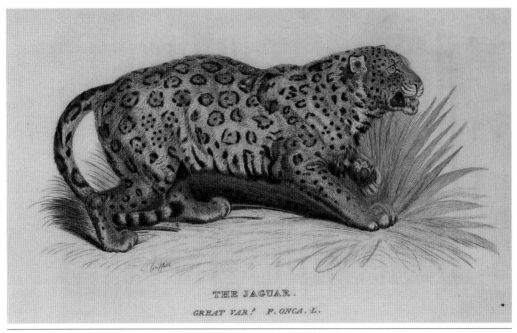

Figure 1. The Jaguar. Great Var F. Onca. L. Etching. (Source: The animal kingdom arranged in conformity with its organization, by the Baron Cuvier with additional descriptions of all the species hitherto named, and of many not before noticed, by Edward Griffith and others. Publisher London, Printed for G. B. Whittaker, 1827–1832)

Figure 2. The Jaguar. Small or Common Var. Etching. (Source: The animal kingdom arranged in conformity with its organization, by the Baron Cuvier with additional descriptions of all the species hitherto named, and of many not before noticed, by Edward Griffith and others. Publisher London, Printed for G. B. Whittaker, 1827–1832)

Figure 3. The Panther of the Ancients. Etching. (Source: The animal kingdom arranged in conformity with its organization, by the Baron Cuvier with additional descriptions of all the species hitherto named, and of many not before noticed, by Edward Griffith and others. Publisher London, Printed for G. B. Whittaker, 1827–1832)

Figure 4. The Panther: *Felis Pardus.* Etching. (Source: The animal kingdom arranged in conformity with its organization, by the Baron Cuvier with additional descriptions of all the species hitherto named, and of many not before noticed, by Edward Griffith and others. Publisher London, Printed for G. B. Whittaker, 1827–1832)

the animals for the broader audience, for whom the color plates were accessible and intriguing, without exploiting the animal body as a symbol of nature, "red in tooth and claw."[61]

NATURALISTS IN THE FIELD

It was very evident to a number of naturalists that direct observation in the field would facilitate opportunities to better inform species identification and histories. Both Alexander von Humboldt (*Personal Narrative of Travels to the Equinoctial Regions of the New Continent*, 1808) and Darwin (*The Voyage of the Beagle*, 1839) published travel journals accounting their journeys in the Western Hemisphere. Both men refer to jaguars of South America in their narratives, but make no mention of the jaguars in Mexico. As late as 1950, the identity, distribution, biology, and ecology of a number of Mexican wild animal species were surprisingly poorly known, due to a lack of recorded observations of species.[62] Prominent naturalists such as Ernest Thompson Seton and John James Audubon dedicated portions of their lives to following the trail of the jaguar, not with the intention of killing a jaguar but for greater insight into the species.[63] These accounts, while physically describing the animal, were also much more in the vein of von Humboldt and Darwin in their colorful descriptions of the animals and their behavior in the environment.

John James Audubon, an ornithologist, naturalist, hunter, and artist, was well known for the quality of his descriptions and detailed paintings of avian species, publishing *Birds of America* in 1827. *The Viviparous Quadrupeds of North America* (1845–1848) was his final work, done in partnership with John Bachman and published by his son posthumously, with Bachman supplying much of the scientific text. The volume was based on observations made in the field in their travels across the continent. These volumes foreground the jaguar-as-hunter, reflecting notions of the time that predators were ferocious beasts to be feared and vilified and leading Lisa Mighetto to comment that "the idea of the predator, then, has been more terrifying than the habits of the animal warrant."[64] Audubon's section on the jaguar begins,

> Alike beautiful and ferocious, the jaguar is of all American animals unquestionably the most to be dreaded, on account of its combined strength, activity and courage, which not only give it a vast physical power over other wild creatures, but enable it frequently to destroy man.[65]

The volume continues with a number of other references to jaguars sneaking up and attacking people. In fact, every mention of the jaguar includes reference to their threat to humans. Another section includes a florid description of a jaguar making a kill:

> This savage beast exhibits great patience and perseverance remaining for hours crouched down with head depressed and still as death. . . . The unsuspecting creature draws near the dangerous spot suddenly with a tremendous leap the jaguar pounces on him and with the fury of an incarnate fiend fastens upon his neck with his terrible teeth whilst his formidable claws are struck deep into his back and flanks. The poor victim writhes and plunges with fright and pain and makes violent efforts to shake off the foe but in a few moments is unable longer to struggle and yields with a last despairing cry to his fate The jaguar begins to devour him while yet alive and growls and roars over his prey until his hunger is appeased.[66]

Audubon's illustration of a jaguar is very different than Cuvier's, reflecting the tone of the text (see figure 5). Teeth and claws bared, the cat is menacing and mere seconds away from attacking. Here, the jaguar is abstracted and condensed into a caricature of its own predatory nature. The jaguar is overtly present, stalking humans behind every tree and bush. These accounts in particular are more tailored to the contemporary perception of predators rather than observed or recorded behavior, as there are very few accounts of jaguars ever attacking people.

Later naturalist narratives more explicitly connected the jaguar as a symbol of the rapidly disappearing wilderness and people's connection to the wild. Perhaps most succinctly, American ecologist, hunter, and nature writer Aldo Leopold shares his reminiscence of the jaguar in his *A Sand County Almanac* (1949), written on a three-week trip to the Colorado River delta with his brother, Carl:[67]

We saw neither hide nor hair of him, but his personality pervaded the wilderness; no living beast forgot his potential presence, for the price of unwariness was death. No deer rounded a bush, or stopped to nibble pods under a mesquite tree, without a premonitory sniff for *el tigre*. No campfire died without talk of him. No dog curled up for the night, save at his master's feet; he needed no telling that the king of cats still ruled the night; that those massive paws could fell an ox, those jaws shear off bones like a guillotine. By this time the Delta has probably been made safe for cows and forever dull for adventuring hunters.[68]

Figure 5. The Jaguar. Lithograph "Drawn from Nature by J. W. Audubon." (Source: The Viviparous Quadrupeds of North America, Published: New York 1848)

This passage evokes many of the images of both the real and imagined landscapes associated with the jaguar. In these narratives, the jaguar is not present, and yet the idea of the fierce man-eater remains a provocative image that "pervades" the landscape. Leopold captures the lived reality and collective imaginings of wilderness as they mingle together in the form of the spotted cat, *el tigre*.

HUNTING NARRATIVES

Although Mexico never gained the popularity of East Africa, India, or Alaska as a hunting destination, for a short time among a select group of sportsman, Mexico became the location of choice for a hunting excursion.[69] Sport hunting, or hunting game as an avocation rather than out of necessity, did not become popular in the United States until the twentieth century.[70] Prior to this time, the killing of predators, including jaguars, was done by ranchers and homesteaders in an attempt to protect livestock. In the late nineteenth and early twentieth centuries, antipredator sentiment was rampant throughout the country, reinforced through federal policy (such as the Predatory Animal and Rodent Control Act of 1915), popular literature that sympathized with "innocent" prey species and demonized predators, and even the rhetoric of President Theodore Roosevelt, who denounced predator species like the wolf for their "vicious, bloodthirsty" natures.[71]

In the U.S.-Mexico borderlands, Brown and Lopez Gonzalez observe, "mountain lions were the usual quarry and could be hunted year-round, [but] every houndsman dreamed of someday catching a jaguar."[72] While shooting a jaguar north of the border was legal at that time, due to the scarcity of the species it was a rare event. Brown and Lopez Gonzalez note that "with rare exceptions, the only people who have had any contact with borderland jaguars during the last century have been hunters and ranchmen. To these people . . . to appreciate a jaguar has always been to kill it. . . . Killing a tigre is, after all, the highlight of a man's life— a trait we found shared by a small cadre of countrymen on both sides of the border."[73]

Jody Emel argues that killing of large carnivores, such as big cats, is only partially motivated by economic reasons; it has "an intertwining causality stemming from a dominant construction of masculinity that is predicated upon mastery and control through the hunt."[74] Humans kill large carnivores for myriad reasons, including the belief that they are a threat to human safety, as a result of depredations of livestock and wild game, for their pelts and body parts, for prestige, and out of animosity and intolerance.[75] However, while they are cast as villains, large predators, including jaguars, are also revered as symbols of masculine power and the ultimate expression of fierce nature, an "indicator of prowess and virility," and taking the life of a large carnivore is an expression of that power, a method of defying fear and possessing the power of the cat.[76] These large carnivores became targets for hatred and outlets for masculine expression, as well as markers of social class and prestige.

It is also important to consider the significance of the social experience of the hunting party and so-called campfire culture in the construction of jaguar narratives. These jaguar-hunting expeditions were not simply naked expressions of masculine bloodlust. They were also an opportunity for a team of men to "take on" the wilderness. Camaraderie and brotherhood are shared in the difficulties and dangers of the trek, the mastery of the kill, and remembrances after the hunt has ended. Around the campfire, stories are woven and rewoven, growing into legend. This was an opportunity for men to claim and enact rituals of the mythic frontier identity so cherished in America.

Traveling into jaguar-occupied areas of Mexico meant sparing great expense to "outfit an expedition in regions with primitive transportation and securing letters of introduction, permits and fees in the hopes of having a welcome reception."[77] Enterprising Americans set up hunting outfits, catering to "avid sportsmen" seeking animals one could no longer hunt in the United States. These "miniexpeditions" brought the rich and famous into the region, such as Gary Cooper and Clark Gable, to shoot and kill a jaguar.[78] The presence of very public figures participating in these hunts further bolstered tone of privilege, excess, and masculinity during these hunts.

There are a handful of narratives written by hunters searching for the jaguar in Mexico. Typically, male Anglos of privilege, who could afford to outfit a team to travel into Mexico in search of the elusive cat, wrote hunting narratives. Other narratives relate tales of enterprising Americans who ran the hunting outfits in this region, perhaps most notably the Lee family of the Tucson, Arizona, area, who started leading jaguar hunts into Mexico in 1935.[79] These stories frequently feature local guides, as well as American Indian assistants, on the trips. These hunting expeditions became a site where Anglos, Mexicans, and American Indians encountered one another, sharing the common experience of the hunt. During the hunt, these men would exchange both myth and knowledge, if not the physical experience, of the jaguar.

These hunting narratives are frequently filled with anecdotes about one's fellow travelers and guides, and, interestingly, most often the jaguar is hardly present in the stories. Rather, these writings reflect the experience traveling through rough country, meeting local guides, culinary adventures, and the thrill of the chase. The emphasis is on working with other men, horses, and dogs, while the cat remains a ghostly specter that legitimizes the purpose of the hunting party.

Frequently, the hunters cannot name their guides and local assistants. While one hunter, Dale Lee, refers to all non-Anglos as "Mexican Boys" in his narrative, each of his hounds is featured by name in the tales, which largely celebrate the dog's prowess in tracking and treeing jaguars. This privileging of the hounds is a common theme shared by these hunting narratives. Another hunter, Frank Hibben, "affectionately dedicated" his 1948 book, *Hunting American Lions*, to the "immortals of the trail," his hounds, each listed by name.[80] The hounds become almost human in some of the tales; as Hibben anthropomorphizes, the "dogs around us pricked up their ears and stopped a second as though they said to one another 'old Jim has found a track in the place and at night? It seems impossible.' But the hounds went to see for themselves in that same fringe of bushes. In a matter of seconds there were three of four other dog voices added to the noise that rolled out over the moonlit landscape."[81]

While the hounds are relatable and almost human, the jaguar is imagined alternatively as a trophy or the embodiment of evil, but never just as the animal, as revealed in hunter Phillip Russell's statement, "All this time we had been impatient to start on a jaguar hunt and even had dreams of carrying home the hide of old Red Paws himself."[82] Clearly, the jaguar was also imagined as a symbol of fierce nature and a challenge to male dominance and superiority. Hibben states, "Most of our conversations took the form of projected trips down into Old Mexico to get on the trail of these devil-cats." Later Hibben fantasizes about what it would be like to encounter the jaguar on the trail at night, "In that brilliant moonlight the yellow and black of his spectacular hide would look like a nightmare and if it didn't, his green eyes certainly would."[83]

In these accounts, local American Indian or Mexican guides are accorded respect for their knowledge of both tracking and jaguar appearance and behavior; they are simultaneously demeaned and portrayed in patronizing terms as simple and childlike. Phillip Russell, while on the trail of jaguars, speaks of his American Indian hosts: "Always they were serene, good-humored and unwearied; took us everywhere and showed us everything; refrained from laughing at our mistakes; and acted as though we were the efficient hunters and fishermen which we were not."[84] However, Russell's tone quickly turns more indicative of perceived power relationships. After describing his handy Swiss pocketknife, Russell notes: "We never sat down to a cocoanut feast where Loretto (the indigenous guide) did not ask for this knife. He loved it. . . . Eventually I gave it to him. He protested that he could not accept it . . .; but his refusal was somewhat weak and finally he pocketed it, saying only gracias. . . . He also called me maestro, master, which was flattering and was pay enough."[85] Through these narratives, the behavior and responses of these guides appear tinged with irony and humor. These men were experienced in playing to what the Anglos expected to receive and played the part accordingly. Russell's entire account of his trip with his Mayan guide is colored with this humor, although it does seem the author is himself oblivious.

Typically, these accounts do not savor in the bloodlust or death of the animal, and these details are often glossed over. Prominent hunting guide Dale Lee recalled, "By golly now, the jaguar charges out and this old boy says 'bang' and down she went."[86] The emphasis was on sport and fair chase rather than on the act of killing. These animals appear as only objects, or foils, but never seem to earn the respect one might expect. Instead, these narratives seem to overlook the jaguar's animality altogether. The jaguar frequently figures into these narratives not as an animal, but imagined as a cunning villain or worthy foe, and as the hunters move across the challenging physical landscape, they also traverse the imagined.

CONCLUSION

Even a brief glimpse into representations of the jaguar in the nineteenth and twentieth centuries reveals a complex, conflicting, and contested set of narratives and images. These naturalist accounts, artistic renderings, and hunting narratives demonstrate the multiple ways in which humans sought to understand this species. One of the most compelling factors about this species is its relative scarcity and absence. In that absence, these representations have a disproportionate impact, as these few observations and narratives of encounter have to speak for the entirety of the species. As a result, some misconceptions, such as the ferocity of the jaguar, would remain uncorrected for nearly 100 years, for while jaguars very rarely attack humans, the legend of accounts such as the one published by Audubon have served as the animal's proxy in its absence.

From confusion regarding its name, physical description, and area of residence to the inflammatory and very likely fictionalized naturalist narratives, jaguar-the-animal has had a difficult time being understood in American and European societies. This legacy lives on today, as very nearly every contemporary issue surrounding the species' home range, genetic classification, and legal status continues to be highly contested and debated. To be sure, the jaguar has become one of the nation's rarest, most misunderstood, and most controversial species.

NOTES

1. David E. Brown and Carlos A. Lopez Gonzalez, *Borderland Jaguars* (Salt Lake City: University of Utah Press, 2001), 67.

2. Elizabeth P. Benson, "The Lord, the Ruler: Jaguar Symbolism in the Americas," in *Icons of Power: Feline Symbolism in the Americas*, ed. Nicholas Saunders (New York: Routledge, 1998), 53–76.

3. Nicholas J. Saunders, "Predators of Culture: Jaguar Symbolism and Mesoamerican Elites," *World Archaeology* 26, no. 1 (1994): 104–117; Elizabeth P. Benson, ed., *The Cult of the Feline* (Washington, D.C.: Dumbarton Oaks Research Library and Collections, 1970); Nicholas Saunders, ed., *Icons of Power: Feline Symbolism in the Americas* (New York: Routledge, 1998).

4. Benson, *Cult of the Feline*. See Peter Boomgaard, *Frontiers of Fear: Tigers and People in the Malay World, 1600–1950* (New Haven, Conn.: Yale University Press, 2001), x; and Jon T. Coleman, *Vicious: Wolves and Men in America* (New Haven, Conn.: Yale University Press, 2004).

5. Michael Woods, "Fantastic Mr. Fox? Representing Animals in the Hunting Debate," in *Animal Spaces, Beastly Places: New Geographies of Human-Animal Relations*, ed. Chris Philo and Chris Wilbert (New York: Routledge, 2000), 201.

6. Steve Baker, *Picturing the Beast: Animals, Identity, and Representation* (Chicago: University of Illinois Press, 2001), xvi; Woods, "Fantastic Mr. Fox?" 182–184.

7. Bruno Latour, *Science in Action: How to Follow Scientists and Engineers through Society* (Cambridge, Mass.: Harvard University Press, 1987), 226–230.

8. Woods, "Fantastic Mr. Fox?" 183.

9. Woods, "Fantastic Mr. Fox?" 183.

10. Woods, "Fantastic Mr. Fox?" 183.

11. For example, see Latour, *Science in Action*; Donna Haraway, *Simians, Cyborgs, and Women: The Reinvention of Women* (New York: Routledge, 1991); David Bloor, *Knowledge and Social Imagery* (London: Routledge, 1976); G. N. Gilbert and M. Mulkay, *Opening Pandora's Box: A Sociological Analysis of Scientists' Discourse* (Cambridge: Cambridge University Press, 1984); A. Pickering, *Constructing Quarks: A Sociological History of Particle Physics* (Chicago: University of Chicago Press, 1984); R. Williams and D. Edge, "The Social Shaping of Technology," *Research Policy* 25 (1996): 856–899; Charles Arthur Willard, *Liberalism and the Problem of Knowledge: A New Rhetoric for Modern Democracy* (Chicago: University of Chicago Press, 1996).

12. Woods, "Fantastic Mr. Fox?" 182–202; Baker, *Picturing the Beast*.

13. R. M. Nowak, *Walker's Mammals of the World*, 6th ed. (Baltimore: Johns Hopkins University Press, 1999).

14. Brown and Lopez Gonzalez, *Borderland Jaguars*, 17.

15. Brown and Lopez Gonzalez, *Borderland Jaguars*, 17; Terry B. Johnson, William E. Van Pelt, and James N. Stuart, *Jaguar Conservation Assessment for Arizona, New Mexico, and Northern Mexico* (Phoenix: Arizona Game and Fish Department, 2009); Nowak, *Walker's Mammals of the World*.

16. Brown and Lopez Gonzalez, *Borderland Jaguars*, 21.

17. James R. Hatten, Anna Laura Averill-Murray, and William van Pelt, "A Spatial Model of Potential Jaguar Habitat in Arizona," *Journal of Wildlife Management* 69, no. 3 (2005): 1024–1033.

18. Hatten, Averill-Murray, and van Pelt, "Spatial Model of Potential Jaguar Habitat," 1024; see also Alan R. Rabinowitz, "The Present Status of Jaguars (*Panthera onca*) in the Southwestern United States," *Southwestern Naturalist* 44 (1999): 96–100; Brown and Lopez Gonzalez, *Borderland Jaguars*, 44–46.

19. Brown and Lopez Gonzalez, *Borderland Jaguars*, 60; Hatten, Averill-Murray, and van Pelt, "Spatial Model of Potential Jaguar Habitat," 1024.

20. George B. Schaller and Peter Gransden Crawshaw Jr., "Movement Patterns of Jaguar," *Biotropica* 12, no. 3 (1980): 161.

21. Alan R. Rabinowitz and B. G. Nottingham Jr., "Ecology and Behavior of the Jaguar (*Panthera onca*) in Belize, Central America," *Journal of Zoology* 210, no. 1 (1986): 149.

22. Hatten, Averill-Murray, and van Pelt, "Spatial Model of Potential Jaguar Habitat," 1024; see also K. L. Seymour, "*Panthera onca*," *Mammalian Species* 340 (1989): 1–9; Richard Perry, *The World of the Jaguar* (New York: Taplinger, 1970), 31–100.

23. Brown and Lopez Gonzalez, *Borderland Jaguars*, 50, 88.

24. Emil B. McCain and Jack L. Childs, "Evidence of Resident Jaguars (*Panthera onca*) in the Southwestern United States and the Implications for Conservation," *Journal of Mammalogy* 89, no. 1 (2008): 1–10; Eric W. Sanderson, K. H. Redford, C. L. B. Chetkiewicz, et al., "Planning to Save a Species: The Jaguar as a Model," *Conservation Biology* 16, no. 1 (2002): 58–72.

25. Bjorn Kurtén and Elaine Anderson, *Pleistocene Mammals* (New York: Columbia University Press, 1980); Steve Pavlik, "Rohonas and Spotted Lions: The Historical and Cultural Occurrence of the Jaguar, *Panthera onca*, among the Native Tribes of the American Southwest," *Wicazo Sa Review* 18, no. 1 (2003): 159.

26. John Lawson, *History of North Carolina* (London 1709), quoted in Peter Matthiessen, *Wildlife in America* (New York: Viking Press, 1959), 42; Brown and Lopez Gonzalez, *Borderland Jaguars*, 32.

27. Matthiessen, *Wildlife in America*, 42; Brown and Lopez Gonzalez, *Borderland Jaguars*, 32–43.

28. McCain and Childs, "Evidence of Resident Jaguars," 1.

29. Brown and Lopez Gonzalez, *Borderland Jaguars*, 52; Perry, *World of the Jaguar*.

30. Brown and Lopez Gonzalez, *Borderland Jaguars*; McCain and Childs, "Evidence of Resident Jaguars," 1.

31. J. M. Philips, "Transplanting the Jungle King," *In the Open*, 4, no. 4 (1913): 13–29, cited in Brown and Lopez Gonzalez, *Borderland Jaguars*, 50; Brown and Lopez Gonzalez, *Borderland Jaguars*, 4; Richard Perry, *World of the Jaguar*, 28.

32. Brown and Lopez Gonzalez, *Borderland Jaguars*, 4.

33. John James Audubon and John Bachman, *The Quadrupeds of North America*, vol. 3 (New York, 1854), 3–6; Brown and Lopez Gonzalez, *Borderland Jaguars*; Neil B. Carmony, *Onza! The Hunt for a Legendary Cat* (Silver City, N.M.: High Lonesome Books, 1995), 14.

34. Nicholas Saunders, "Architecture of Symbolism: The Feline Image," in *Icons of Power: Feline Symbolism in the Americas*, ed. Nicholas Saunders (New York: Routledge, 1998), 14–15.

35. Pavlik, "Rohonas and Spotted Lions," 158–175.

36. Karl Luckert, *The Navajo Hunter Tradition* (Tucson: University of Arizona Press, 1975), cited in Pavlik, "Rohonas and Spotted Lions," 165.

37. Pavlik, "Rohonas and Spotted Lions," 165.

38. Saunders, "Architecture of Symbolism," 14–15.

39. Saunders, "Architecture of Symbolism," 14–15.

40. Brown and Lopez Gonzalez, *Borderland Jaguars*, 4.

41. Ernesto Alvarado Reyes, "The Legend of the Mexican Onza," *Mastozoologia Neotropical* 15, no. 1 (2008): 147–148; Carmony, *Onza!* 12–14.

42. Reyes, "Legend of the Mexican Onza," 147–148.

43. A. Starker Leopold, *Wildlife in Mexico: The Game Birds and Mammals* (Berkeley: University of California Press, 1959); Reyes, "Legend of the Mexican Onza," 147–148.

44. Carmony, *Onza!* 12.

45. Carmony, *Onza!* 8.

46. M. J. P. Currier, "Felis concolor," *Mammalian Species* 200 (1983): 1–7.

47. Brown and Lopez Gonzalez, *Borderland Jaguars*, 15.

48. Carmony, *Onza!* 12.

49. Carmony, *Onza!* 12.

50. Brown and Lopez Gonzalez, *Borderland Jaguars*, 17.

51. Brown and Lopez Gonzalez, *Borderland Jaguars*, 15.

52. Henry Lowood, "The New World and the European Catalog of Nature," in *America in European Consciousness 1493–1750* (Chapel Hill: University of North Carolina Press, 1995), 295–323.

53. Georges Cuvier, Le regne animal distribute d'apres son organization (Paris: Chez Deterville, 1817); Georges Cuvier, *The Animal Kingdom Arranged in Conformity with Its Organization*, ed. by Edward Griffith (London: Whittaker and Company, 1827).

54. Cuvier, *Animal Kingdom*, 114–116.

55. Cuvier, *Animal Kingdom*, 114–116.

56. Cuvier, *Animal Kingdom*, 114–116.

57. Cuvier, *Animal Kingdom*, 114–116.

58. *The Penny Cyclopedia of the Society of Useful Knowledge*, vol. 13 (London: Charles Knight, 1839), 430–435.

59. *Penny Cyclopedia*, 430–435.

60. K. P. Bland, "Tom-cat Odour and Other Pheromones in Feline Reproduction," *Veterinary Research Communications* 3, no. 1 (1979).

61. Stephen Jay Gould, "The Tooth and Claw Centennial," in *Dinosaur in a Haystack: Reflections in Natural History* (New York: Three Rivers Press, 1996).

62. Carmony, *Onza!* 15; Leopold, *Wildlife in Mexico*.

63. Brown and Lopez Gonzalez, *Borderland Jaguars*, 6.

64. Lisa Mighetto, *Wild Animals and American Environmental Ethics* (Tucson: University of Arizona Press, 1991), 81–82.

65. Audubon and Bachman, *Quadrupeds of North America*, 3–6.

66. Audubon and Bachman, *Quadrupeds of North America*, 3–6.

67. Aldo Leopold, *A Sand County Almanac (*Oxford, U.K.: Oxford University Press, 1949); Neil B. Carmony and David E Brown, *Mexican Game Trails: Americans Afield in Old Mexico, 1866–1940* (Norman: University of Oklahoma Press, 1991), 148.

68. Leopold, *Sand County Almanac*, 143–144.

69. Carmony and Brown, *Mexican Game Trails*, 3.

70. Brown and Lopez Gonzalez, *Borderland Jaguars*, 96.

71. Mighetto, *Wild Animals and American Environmental Ethics*, 75–93; Theodore Roosevelt, "Hunting the Grisly and Other Sketches," in *The Works of Theodore Roosevelt*, vol. 13 (New York: P. F. Collier and Son, 1893), 225.

72. Brown and Lopez Gonzalez, *Borderland Jaguars*, 97.

73. Brown and Lopez Gonzalez, *Borderland Jaguars*, vii–4.

74. Jody Emel, "Are You Man Enough, Big and Bad Enough? Wolf Eradication in the U.S.," in *Animal Geographies: Place, Politics and Identity in the Nature-Culture Borderlands*, ed. J. Wolch and J. Emel (New York: Verso, 1998), 91–118.

75. Emel, "Are You Man Enough," 91–118.

76. Emel, "Are You Man Enough," 91–118.

77. Carmony and Brown, *Mexican Game Trails*, 3.

78. Carmony, *Onza!* 69, 78.

79. Carmony, *Onza!* 53–105.

80. Frank C. Hibben, *Hunting American Lions* (New York: Thomas Y. Crowell, 1948).

81. Hibben, *Hunting American Lions*, 160.

82. Carmony and Brown, *Mexican Game Trails*, 211.

83. Hibben, *Hunting American Lions*, 165.

84. Carmony and Brown, *Mexican Game Trails*, 211.

85. Carmony and Brown, *Mexican Game Trails*, 216.

86. Carmony and Brown, *Mexican Game Trails*, 243.

Animal Deaths and the Written Record of History
The Politics of Pet Obituaries

JANE DESMOND

A BLACK LABRADOR DOG NAMED BEAR BROKE THE SPECIES BARRIER AND MADE IT into the obituary pages of my local paper a few years ago (see figure 1). He was the first animal ever so commemorated in that newspaper (the *Iowa City [Iowa] Press-Citizen*). Bear died quietly in his sleep on the morning of April 14, 2003, at thirteen years of age. But if his passing was peaceable, the period after his death was not. His obituary became the cause of bitter debate in our community. Why would a simple publication in a local newspaper generate such a firestorm of emotion? What is at stake?

In this chapter, I analyze contemporary U.S. debates over publishing pet obituaries in newspapers. I suggest that a pet obituary raises multiple issues about the "appropriate" objects of mourning, of the "right" to mourn publicly, and of the ways that public mourning legitimates social relationships. I examine how obituaries for animals stretch the obituary form and push the limits of writing lives into the social record. Finally, I ask: what phenomena of social value are being contested in these debates? I argue that pet obituaries are seen as dangerous by some precisely because they are a political act signaling the need for far-reaching ethical change.

To begin to explore these issues, let me return to that obituary of Bear, the dog. Bear's obituary was placed on the page next to that of Georgi Addis, whose sister-in-law, Sue Dayton, became incensed. She found the inclusion of the dog's obituary on the same page "distasteful and disrespectful to Georgi and [her] family," and in a letter to the editor called on the paper to issue an apology to every family involved with the obits printed that day. Ensuing letters to the editor argued pro and con.[1]

A similar debate erupted right around the same time in Philadelphia when the *Philadelphia Daily News*

Bear, 13

Bear was known widely in the Iowa City community for his gentle humor and his tendency to nap on the streets.

An inveterate lover of humankind, and particularly of the ladies, he was at home wherever there was an outstretched hand, and cherished those  **Bear** most whose offerings were most dependable.

Bear passed in his sleep on the morning of April 14, at 13 years of age.

Figure 1. An obituary for Bear, the black Labrador, age thirteen. (Source: *Iowa City [Iowa] Press-Citizen*, April 15, 2003)

printed an obituary for Winnie, a nine-year-old terrier known for her "feisty and fearless" nature. That obit ignited national discussion in columns in the *Washington Post* and in *Media-Week*, where writer Lewis Grossberger termed it a "grotesque phenomenon."[2]

These current contestations over the public newspaper memorialization of pets chart both the shifting status of the "pet" in terms of family and kinship relations and the changing practices in obituary writing as authority moves from the funeral director to the family in the era of paid obituaries.

It appears that the concept of a pet as a "part of the family" is being taken in more literal ways than previously. A few statistics indicate the scale of this issue. The American Pet Products Manufacturers Association reports that 34 percent of U.S. households own at least one cat, and 39 percent own at least one dog, for a total of 75 million dogs and 88 million cats living in U.S. households in 2008. A poll based on a random sample of 1,000 pet owners and taken by the Associate Press in 2009 reveals that 50 percent of all pet owners surveyed regard their pet as a *full* member of the family (with a low of 43 percent of married males to a high of 66 percent of unmarried females).[3] They agreed with the statement, "My pet is just as much a part of the family as any other person in the household." A further 36 percent also regarded the pet as part of the family "but not as much as the people in the household." Such polls are certainly not conclusive and actually tell us little about exactly what that notion of "full family member" means;[4] however, they suggest a substantial commitment to and emotional investment in the animal.

Pet obituaries articulate an extended notion of kinship obligations and recognition by publicly recognizing this bond with nonhuman animals. Obituary conventions are already stressed by the changing recognition of same-sex partners, but pets push this boundary of acceptable kinship further. Pet obituaries, which publicly commemorate a life and make that life part of the historical record, provide one of the test cases of this shifting positioning of the "pet" in relation to human companions.[5] As with humans, pet obituaries assign value to a life, define its highlights, extol socially validated accomplishments, and serve as models of living.[6]

HISTORY OF OBITUARIES

Like so many other social conventions that we take for granted, U.S. obituaries, as a genre of writing and publication, have a history that has changed over time. Once reserved primarily for wealthy and powerful white men, public recordings of life and death have gradually expanded to encompass a much wider range of the population.[7] While major newspapers like the *New York Times* may still feature the obits of famous people, local papers will print death notices or obituaries of just about anyone. In this sense, death has become democratized. As historian of English-language obituaries Nigel Stark puts it: "It is through the obituary, above all other forms of journalism, that an insight is obtained of what it was like to be a citizen of a particular community at a particular time, for it offers a sustained, often dramatic reflection, of prevailing mores."[8] As such, even the more democratized obituary form still reflects the contours of social hierarchy while expanding the notion of whose death is worthy of public notice.

For example, recent research has found that women continue to receive significantly fewer obits and, even when included, to have shorter obits than men in the same paper.[9] Euro-Americans continue to receive longer and more complex obits than African Americans.[10]

Some of these differences may be attributed to continuing racialized and gendered differentials in social power, prestige, and professional accomplishment, but they also attest to the function of the obit as a measure of presumed social worth.

In the past, there have been two categories of animals whose death were deemed newsworthy: animals of the famous or famous animals. For example, Igloo, Admiral Richard Byrd's dog and a "polar hero" himself, received an obit in the *New York Times* in 1931.[11] Titled "Igloo, Byrd's Dog, Polar Hero, Is Dead," the piece featured a picture of Igloo, a white and black fox terrier, posed with Admiral Byrd, accompanying a long discussion of key points in his life, including the moment when, riding in a plane, he and Byrd circled the North Pole and "he became the most famous dog in the world." Considerable space is given over to describing his personality, noting that "his courage and daring exceeded his [small] physique."[12] Simple funeral rites were scheduled at the Pine Ridge animal cemetery in Dedham, Massachusetts. Although a news article, the piece includes the hallmarks of the obituary style, noting age, place and cause of death, life accomplishments, and grieving survivors, along with funeral plans. A later report on the funeral pictured Byrd, his son, and the pallbearers, including Master Sergeant Victor H. Czegka, a member of the South Pole expedition, who carried the small white casket to the grave, where Byrd placed a spray of white roses.[13]

More recently, Marjan, the one-eyed lion in the Kabul zoo who became a symbol of war-torn Afghanistan, received attention in the *Chicago Tribune* with a 2002 article titled "Kabul Zoo's Shell-shocked Lion Dies." Starting with a description of the "people from all walks of life who came . . . to the battle-scarred Kabul Zoo to pay homage," the article notes the lion's physical ailments, the work of the veterinarians, and the past horrors of his captive life, including being starved, bombed, and beaten. His role as an international symbol of the need to rebuild both the nation and the zoo was emphasized, as were details of his burial.[14] Importantly, these appreciations appeared in the news pages, not the obituary sections, but they still detailed the lives, accomplishments, and personalities of individual animals.

The democratization in obituaries has only recently expanded to encompass the ordinary pets of ordinary people. The national move to charge for obituaries has been key in enabling this shift. By 2002, paid obituaries had become the national norm, although some papers were still making the transition.[15] Paid obits written by the family, and not by the newspaper staff, assure that the notice gets in the paper in a timely manner when there is a backlog on the obit writing staff and also allow families to customize their notices by spending extra to include "photos, symbols and memorial information." This publicity does not come cheap. While most newspapers continue to publish free death notices of a few short lines, paid obits can cost more than $10 a line, with extra for a photo.[16] A ten-inch obituary with a photo would cost $634.14 in the *Dallas Morning News*.[17] "Cash not cachet" may become the new criterion for obituary inches.[18]

Some journalists worry that home-written obits are no longer subject to fact-checking and cannot be considered an official record. More important, perhaps, touchy social issues can invade the paper. For instance, editors will no longer decide whether gay partners should be included as survivors or whether an unborn child who died at twenty-five weeks in gestation could have an obituary. Or, could unsavory sentiments be included? As one media critic worries, "should the angry, long-abused sons of a hard-drinking father be allowed to include that 'dad was a mean drunk' in their copy?"[19] In all pet obituary cases I have found, the pet obits have been paid obits. People are taking advantage of the declining authority of newspaper editors to shift the boundaries of the includable.

It is hard to track down information about exactly how widespread this phenomenon is since the pet obits are not indexed. However, it is clear that the practice, though still in its infancy, is spread throughout the United States and is engaging both major and minor newspapers. Pet and human obits appear side-by-side by chance or by policy, even sometimes by subterfuge.

The papers participating are diverse in size and geographic range. Among them are small community papers like the *Sun* of Bremerton, Washington, a daily with a circulation of 33,000, and the smaller weekly *Jamestown News* in North Carolina.[20] But the phenomenon is not limited to newspapers with tiny circulations. The larger *Anchorage (Alaska) Daily News* published the obituary of Carl, also known as Mr. Handsome, the dog of the former governor of Alaska, in the regular obit section in August 1999. A newspaper spokesman called this "an unfortunate mistake," saying the newspaper's policy was to run pet obits anywhere but on the human obits page.[21]

Another "mistake" occurred in the 1990s when the obituary editor of the *Durham (N.C.) Morning Herald* received a mysterious phone call from a local funeral director. He wanted a death notice (a short two-inch announcement) put in the obituary section the next day, and gave the name as a human first name and his own last name. He refused to give the age, but noted when the death occurred. The dead "person" turned out to be a dog, as the newspaper found out the next day when a local citizen who knew the dog's family called to express his outrage that a dog was on the obits page with the humans. The editor was furious about being tricked, but the example also shows how savvy people can work the system (he was a funeral director after all) to stretch its limits, improvising a space for pet mourning when no such official one exists.[22]

Like Mr. Handsome in Alaska, Axel, a police dog, also got special treatment because he himself was special. In 2001, the *News Tribune* of Tacoma, Washington, published his obituary complete with a photo of him posed with his police partner, Officer Stephen Shepard, with the tag line "Tracking bad guys in Heaven" running under the picture. His courage in apprehending 220 felons was praised in improving the safety of the lives of Tacoma's citizens. The obituary was printed on the regular obits page right next to the obit for Earl Criss, a man whose list of survivors included "his beloved dog, Maggie." This one page thus demonstrates two moments of disruption in obituary practices, both the inclusion of animal obituaries among human ones and the new trend of listing pets among a deceased person's grieving family members, exemplifying the expanding notion of the family across species.[23]

PUBLIC CONTESTATION

These examples of pet obituaries might have remained isolated local incidents rather than indicators of a national phenomenon, but when the *Philadelphia Daily News* announced in 2002 that it was going to start printing pet obituaries (actually pet death remembrances to be published once a month in a special section in the classified ads titled "A Fond Farewell to Our Beloved Pet"),[24] the Associated Press picked up the story, and a media blitz exploded. As the *American Journalism Review* reports: "CBS News called, as did NBC's 'Today' show. Local TV and radio in Philadelphia reported it; 'Imus in the Morning' talked about it; Canadian public radio ran a story; Robert Siegel of NPR's 'All Things Considered' covered it; and CNN's 'Talk–Back Live' devoted two discussion segments to it."[25] Bob Levey's "Washington" column in the *Washington Post* headlined it the "Latest [in] Questionable Taste,"[26] and Lewis Grossberger skewered the idea in his scathing satirical column in *MediaWeek* titled "Spot Dead! City Mourns."[27]

The alarmist hyperbole caricatures the obits, with Bob Levey saying the implications "are too awful to contemplate. . . . We will soon be buried in a tidal wave of mawkish cuteness. . . . Dogs, cats, hamsters, ponies, iguanas and macaws will take over the obit page, crowding out your poor Uncle Oscar who was not so photogenic as Skippy the distinguished cocker spaniel, who suddenly took ill after ingesting an excessive amount of dead squirrel and antifreeze, having, unfortunately, slipped his chain." Or, he declaims, "we will be informed that former canine officer Blitz, who expired at his retirement kennel in Florida, was a graduate of the Centurion Attack-Dog Academy, where he was voted Most Likely to Seize a Fleeing Perpetrator by the Throat." Or we will "note that funeral services for Edwin the late, beloved, goldfish, probably father of entire schools of descendants, will be held tomorrow in the toilet of his bereaved owners."[28]

Beneath the venomous humor several real issues as well as clear moral judgments emerge. Clearly, for Bob Levey, mourning animal death can only be a mockery of the valor and dignity of human lives. The hierarchy of value is clear, and to challenge it is to descend into the realm of the ridiculous, or worse, the embarrassment of "bad taste," here feminized as excessively mawkish and emotional.

And second, just what would a "proper" animal obituary consist of anyway? Obits are a literary genre—they report birth, parents, education, marriage, divorce, children, grandchildren, work, and membership in fraternal, religious, or service organizations, along with hobbies and moral or personality traits. So how must the genre change to accommodate animals? What ideals of "petness" do we underline in the obits of pets? Should there be an emphasis on "immanent" personality traits (loyalty, generosity) or on accomplishments (won best in show, or saved a child from drowning)? How do we list survivors? Do we include other animals? The form is straining under these challenges, and samples have varied from short appreciations of the "we will miss you brave dog" type to long, wrenching cries of loss that detail the owner/guardian's anguish.

Columnist Betty Cuniberti of the *St. Louis Post-Dispatch* (quoted in Bob Levey's column) also fears the embarrassment of bad taste. Discussing the printing of pet obits among human ones, she says, "I'm trying to imagine a sorrowful son opening our newspaper to look for his mother's obituary and finding her picture next to one of a hamster." Even worse, Levey summarizes, would be "the issue of the hamster's many survivors." Cuniberti imagined there would be 427 hamster children and grandchildren, "including 154 named Fluffy."[29]

The choice of the hamster, easier to satirize than a golden retriever or the cat who has been with a family for twenty years, tries to reduce the pet obit issue to a dismissible absurdity. Its premise is based not only on the differential attribution of value to human and animal life but also to gradations of value among animal species.[30] Ultimately, what is at stake is the "insult" to human life that, for some, public records of pet death threaten. A parallel form implies a parallel value, a distasteful concept for some, abhorrent for others. This was the source of outrage about Bear's obituary.

This type of media outcry did not occur earlier when pet death was confined to online virtual cemeteries or to online memorial pages or to physical pet cemeteries. Despite their large scale, those modes of pet commemoration have not attracted such venomous resistance, because their "public" consists solely of like-minded individuals. Based on my analysis of

500 online postings, these individuals appear to be mostly women who love their pets. The online memorials, posted on sites like Petloss.com, which houses more than 50,000 obituaries for individual animals, are a sort of "private public" sphere. The public for these sites must intentionally seek them out and thus most likely consists of those who value what those sites enable: a public legitimation of loss, the elicitation of sympathy and understanding, and an online community of people going through the same thing.

Such legitimation of loss, and of the object of the loss as an appropriate object for mourning, is often unavailable in other realms of daily life where, for instance, a pet's death is not seen as a legitimate reason for staying home from work. A common response to someone announcing the death of his or her cat, for example, might be: "Oh, too bad. Are you getting another one?"[31] Like a broken toaster, the pet can be replaced by another in the same commodity category, in this case feline. Online memorials work against this genericization by mourning a particular feline, and legitimate both the grief and the griever, the latter of whom becomes cast in the role of "survivor," not merely of "owner," and thus is acknowledged as worthy of emotional caretaking.[32]

Newspaper obituaries are fundamentally different. They are there in the historical community record for anyone to see. They do not have to be sought out specifically but rather land on our table, in the news pages flopping open by the bacon and eggs, inserting themselves into every household. Potentially engaging readers of all persuasions, they are seen by many not as an appropriate extension of the obituary form but rather as an explicit challenge to it, a desecration or defrauding of the form.

Pet obits also assert the legitimacy of a pet's family member status, and implicitly legitimate mourning for that familial loss. This challenges the constitution of the "family," a concept already under stress from shifting configurations ranging from single-parent households to gay families. The decision in October 2002 of the *New York Times* to include gay civil ceremonies in its marriage announcement pages exemplifies this tension and the centrality of the public record as legitimator, and similarly redefines kinship.[33]

The most common solution so far has been for newspapers to put pet obits as paid announcements in special sections of the classified ads. This is the move that my local paper announced after receiving letters to the editor decrying the inclusion of Bear's obituary in the regular (human) obits section. This solution—paid announcements in the classifieds—manages the tension between animal as subject and animal as object by granting an obit but grouping it with property. Even so, this in-between status drives a wedge into the convention of obits—implicitly rendering an ordinary animal's death and life worthy of public acknowledgment by writing the individual lives of household pets into the historical record.

But despite these very public debates, people still persist in writing obituaries for their animals and trying to publish them in the obituary pages. The most recent example I have come across is from the *Santa Barbara (Calif.) Independent*, a substantial weekly running about 100 pages per issue. On October 30, 2008, it ran an obituary of Cyrus the cat, "May 1998–August 2008," which featured a photo, just like the three other (human) obituaries on the obituaries page that week. Unlike most obits, however, this one was signed, authored by Vivian Olsen and Viola Freeman, who saluted the ten-year-old long-haired orange tabby with "big eyes, a hairless tail, and golden fur," who was "so intelligent and sweet."[34] I have not yet found any follow-up complaints in the paper, so it may be that weeklies that are seen more as community services than regular "hard" news publications might offer a more welcoming venue for breaking these conventions.

When I discussed these issues with obituary writers at the 2008 meeting of the International Association of Obituarists outside of Albuquerque, New Mexico, strong opinions emerged. For some obituary writers, the issue is mute. Kay Powell, award-winning obituary writer on the staff of the *Atlanta Journal Constitution*, believes that obituaries for animals are oxymoronic. By definition, she says, an obituary is a life story about a person. This opinion comes from the writer who once penned a published obituary for the planet Pluto when it was demoted from major celestial body to the new, less-prestigious category of "dwarf planet" by the International Astronomical Union in 2006. Pluto's "survivors" were listed as Jupiter, Mars, Earth, Mercury, Neptune, Uranus, Saturn, and Venus. But in that instance, although the piece follows the regular obituary form, noting the deceased's influence on Walt Disney, composer Gustav Holst, and astrologists, it was not a "real" obituary. This was all tongue in cheek, sharing in the general sniggering that accompanied the poor planet's demotion in the popular press. When it comes to real obituaries for nonhuman creatures rather than planetary matter, Powell is adamant: "Obits are for people. Pets are animals. Period." For others, the problem of parallelism emerges—a parallel form seems to imply a parallel value, and that is just not acceptable, or even conceivable, for some. It also offers particular challenges for writing.

CHALLENGES OF WRITING A NONHUMAN ANIMAL'S LIFE

As discussions with obituary writers made clear, an obituary is not a story about a death; it is a story about a life. If the inscription on a tombstone is the shortest life story of all, then the obituary comes in second, followed by the expansive forms of a memoir and biography. And these life stories in obituary form have a certain format. Obituary writer and journalist Marylin Johnson describes it this way: "There is . . . a template for the obituary that almost every newspaper publication follows. . . . Writers have absorbed it and readers come to expect it, almost as if an invisible metronome accompanied its unfolding."[35] These required pieces of the template include the "bad news" (the person died) and the "death sentence" (why and how), followed by paragraphs of anecdote often describing a unique turning point in the subject's life, and ending (if the paper is British at least) on a note of elegant, literary irony.[36] The point is to announce the death and emphasize the life story. This is where a particular challenge arises in accommodating the obituary form to encompass animals.

And what can you say about the life story of a dog? The unknown (to humans) interiority of a dog's perception of the world, of his or her actual daily experience, presents a boundary. We cannot know the significance of things in the world, and of relationships, actions, and events, from the dog's point of view by interviewing family members and friends, or from commentary left behind by the individual, because the species difference means that the world is a different world for the dog. That difference means that those markers used by obit writers to craft the life story will not work. This nontranslatability is part of the source of biting humor (see examples above of accomplishments like "graduated from puppy kindergarten" and so on). That the dog, for example, does not, and cannot, fit our categories of social contribution and social worth makes the dog's life appear insignificant and like a failure when measured by these human standards.

To understand this challenge to or incompatibility with the form, we can examine the tips offered in a document meant to help grieving families prepare human obituaries. *Honoring*

Your Loved One in Print, a booklet published by the *News-Gazette* of Champaign, Illinois, states that "through a personalized obituary, you can share thoughts about your loved one by writing about the personality, the relationships, the accomplishments, the interests, and the associations that made up the tapestry of his or her life."[37] The "checklist for creating an obituary" suggests including facts like full name and age; time and place of death; birth date; parents' names; marriage date and name of spouse; survivors, including names of children, siblings, and parents; aspects of life history, including occupation, memberships, interests, activities, and accomplishments; and information about funeral services.[38]

For animals, many of these are unknown (who are the "parents"?) or considered unimportant unless the animal, like a dog, was American Kennel Club (AKC) registered. Other categories, like work, are irrelevant, unless the dog was a "working dog," such as the assistance dogs discussed in Avigdor Edminster's chapter in this volume. Still others, such as "activities," appear ludicrous in human terms. Does your parakeet like to look at himself in the mirror or sing along with music on the radio?

Below is my attempt to engage the challenge of the human obit form when transferred to animals, in the form of an unpublished obituary for my pet rabbit, Baylor. Not only are the categorical distinctions of valuation a challenge to the form, but also there are sensorial differences in how a nonhuman animal, like a rabbit or a dog, experiences the world when compared to humans. They have a fundamentally different way of being in the world. They literally see the world differently, knowing it through smell, and they have a different sense of acting physically in the world. How do we know what the highlights of this life are for that individual being?

An Obituary for Baylor Wabbit

Baylor Wabbit, senior gentleman and shyly gracious fellow, died on Memorial Day weekend, 2008, at the University of Illinois Veterinary College, after what appeared to be a sudden stroke. Excellent emergency care was provided by the small animal vets, but Baylor's age, at 8–1/2, rather advanced for a rabbit, worked against him.

Baylor was born in Iowa and spent his first several years there with a human companion before a brief stint in the Iowa City Animal Care Center, where we found him available for adoption. His silvery coat and dark chocolate–colored eyes, paws, and long floppy ears gave him a sophisticated look that stood out from the crowd there, and caught our eye. Baylor trained us in rabbit care, especially regarding the necessity of fresh vegetables, above all carrots with their leafy tops still attached. These were ideally followed by something sweet, like yogurt drops or a piece of fully ripe banana. The drops were (apparently) so good that they—combined with a special Morse code of taps and entreaties—could even tempt him back into his wire house as he traded footloose freedom for tasty treats.

Baylor established his supremacy in the home at once, and when Mama the dog joined our household soon after his arrival, he stood his ground and did not back away from a sniffing investigation, earning him instant respect from our curious canine. Later he was joined by a second house rabbit, a white floppy eared boy named Christopher, who brought out Baylor's playful side, and stimulated him to hop, dig, and leap around the kitchen where their homes (big wire pens on wheels) took up most of the space. For these two, each was the other's most important companion, although Baylor needed to keep his distance to establish his superiority in the rabbit ranking. Despite this, each often spent the night sleeping by the other, separated only by the widely spaced metal mesh of one cage as the other, on the outside, had roaming rights that day.

Baylor spent five years with us, most in Iowa City, but the last few months in Champaign, Illinois, after a smooth move to a new house that brought him the pleasures of a dedicated rabbit room, with a window for fresh breezes, lots of room for cardboard boxes to deconstruct, and a glass door through which to keep an eye on the workings of the household.

Strong, quiet, gentle, soft, and opinionated, "Mr. B" was a big presence in a small body.

SOCIAL WORTH AND SOCIAL OBLIGATION: WHY PET OBITUARIES ARE DANGEROUS

Beyond the challenge of translating the obituary form from human to animal life is the issue of social value—of social worth—and this is an even more fundamental dividing line. We can certainly imagine human situations where social contribution by such measurements is also minimal: babies, drug addicts, murderers, deeply mentally disabled persons, and even more broadly those whose life contributions revolve around a very small circle, such as family. Do these situations offer us any clues?

A child who dies in infancy after a severe illness may have deeply touched the lives of his or her family circle and even the medical professionals who treated him or her. Perhaps a church or other organization was also involved in caring for the child, praying for him or her, raising money for medical bills, assisting the family, and so on. If so, the reverberations of this individual's life are already wider than that of most domestic pets, moving outward beyond the immediate family unit. But this status, if we can use that word here, is due less to anything the small child might have "accomplished" than to his or her status as a child at the center of widening circles of attachment and care.

Few would imply that such a life is "unworthy" of an obituary or that such a story would be somehow offensive to others whose lives by more conventional measures of accomplishment were listed on the same page. Being a "public figure" can catapult an animal into the newsworthy category with a news item about its death, but social measure of public contribution is not necessary for the death of an individual person to be worthy of an obituary, not a news obit perhaps, but a family paid death notice at least. However, it is precisely these last two forms that are deemed by some to be so offensive when applied to animals.

Social worth becomes a useful measure for understanding attitudes toward obituaries when we apply it not to the individual but to the category. Humans are granted an intrinsic social worth merely by being born into that category of being, whereas animals may earn this membership through notoriety or heroism. And in most cases, this possibility is extended only to pets, not to un-individuated animals denoted as "wild" or as "food."

The real threat of obits for animals is that they are a wedge into the epistemological boundary of the "worthy/not worthy." By pointing to and contesting that boundary through publicly proclaiming a parallelism between the human and nonhuman animal, obits raise the specter of the shifts in social practices that the attribution of such value would portend, including a radical reorganization of food, sports, some scientific research, and so on. No wonder the categorical threat—the threat to supposedly distinct conceptual categories and hence of moral obligations to humans and animals that an animal obit stands for—has been so strongly resisted through the twin techniques of denigration and moral outrage. Writing animal lives into the historical record through obituaries offers the potential for a fundamental challenge to the status of animals and the constitution of the "family" as a social unit.

NOTES

My thanks to those many friends and colleagues who have helped me think about these issues and who sent me clippings of animal obituaries and news stories from around the country as well as copies of those pet obituaries they have written themselves for their own pets and sent to their circle of friends, including Mary Bennett, Susanne Shelton, MacKinze Cook, Shirley Wajda, Kathryn Henry, Joan Adler, and Kathy Cunneen. I thank organizer Carolyn Gilbert of the 2008 International Association of Obituarists conference for inviting me to present my work there and enabling me to get feedback from participants, especially Larkin Bradley, Kay Powell, and Jean Olsen.

1. Bear's obituary appeared in the regular obituary section of the *Iowa City (Iowa) Press-Citizen*, April 15, 2003. Like all the other obituaries, it was titled simply by name and age, in this case "Bear, 13" and accompanied, like many of the others, by a black-and-white photo (showing Bear outdoors lying in the grass staring into the camera). On May 8, 2003, Georgi Addis's sister-in-law Sue Dayton of Iowa City published a letter to the editor in the same paper with the title "Apology Owed for Dog Obituary." Follow-up discussion in the paper included a letter to the editor titled "Have the Editors Lost Their Minds?" by Tami Bryk on April 21, 2003, which included in its lists of several complaints about the newspaper's choice of news stories the statement: "As a pet owner myself, I understand completely the loss of a beloved pet. However, including it with the loss of a 91-year-old woman and her extensive contributions to humanity or the tragic loss of an infant son is simply appalling." The writer calls for a separate section for pets if the paper wants to publish pet obituaries. Finally, in another follow-up on May 10, 2003, in a column by the managing editor, Jim Lewers, "Meet Your Managing Editor," Lewers apologizes for not anticipating the best way to handle pet obituaries when the paper shifted to paid obituaries in 2002. "I apologize for that mistake, particularly to those whose loved ones['] obituaries appeared on the same page as Bear['s]. . . . We've [now] decided to run all other pet obituaries in the classified section." A final act in this public debate occurred in print in another letter to the editor on May 16, 2003, titled "Pet Obituaries Belong in Paper" and signed by several members of the board of directors of the Friends of the Animal Center Foundation (supporting the work of the local animal shelter), including myself. This letter suggested a weekly "pet memorials" section be established in the paper, because placing such announcements in the classified section as the managing editor proposes "reinforces the perception that pets are property." After writing this chapter, I see this as a useful compromise position at the present historical moment, but not as a final solution to the question of how to facilitate public mourning for animals.
2. Lewis Grossberger, "Spot Dead! City Mourns," *MediaWeek* 12, no. 18 (June 5, 2002): 38, accessed through EBSCO database on June 11, 2002.
3. Accessed at http.hsus.org/pets on June 29, 2009, with statistics from the American Pet Products Manufacturers Association 2007–2008 National Pet Owners Survey reported there. For a report on the Associated Press poll, see http://www.petside.com/the-sidewalk/ap_pets_poll.php (accessed June 28, 2009).
4. The Associated Press poll in 2009 gives some indication of a range of actions such a notion of "full part of the family" can indicate to individual humans, including buying gifts for the animal,

giving the animal a humanlike name, letting the animal sleep on the bed, and so on. This does not give us a sense of the hierarchy between human and nonhuman animal members of the family, however.

5. Throughout this chapter I use the term "pet" instead of "companion animal" because it is more widely used in the discourse I am analyzing, and I also use the terms "animal" and "human" instead of "nonhuman animal" and "human animal" to emphasize the conceptual distinction that is always being drawn in these debates about categories of beings deemed worthy of obituaries and those who are not.

6. Pulitzer Prize–winning obituary writer Jim Sheeler makes this point about obituaries as models for living in his book *Obit: Inspiring Stories of Ordinary People Who Led Extraordinary Lives* (New York: Penguin Books, 2007), ix.

7. Janice Hume, *Obituaries in American Culture* (Jackson: University Press of Mississippi, 2000).

8. Nigel Stark, *Life After Death: The Art of the Obituary* (Victoria, Australia: Melbourne University Publishing, 2006), 46. This is a wonderful comparative study of the histories of English-language obituary writing in England, the United States, and Australia.

9. Karol K. Maybury, "Invisible Lives: Women, Men, and Obituaries," *Omega* 32, no. 1 (1995–1996): 27–37; Robin D. Moremem and Cathy Cradduck, "'How Will You Be Remembered After You Die?': Gender Discrimination after Death Twenty Years Later," *Omega* 38, no. 4 (1998–1999): 241–254.

10. Alan Marks and Tommy Piggee, "Obituary Analysis and Describing a Life Lived: The Impact of Race, Gender, Age, and Economic Status," *Omega* 38, no. 1 (1998–1999): 37–57. New studies calculating the effect of the new era of the family-written paid obituary will be needed to determine if "cash not cachet" will have an effect on these social differentials in number and complexity of obituaries.

11. A year later, in 1932, another canine's death received the *New York Times* coverage when movie star Rin Tin Tin died. The photogenic German shepherd had starred in a series of popular adventure movies in the late 1920s and early 1930s. See "Rin Tin Tin Dies at 14 on Eve of 'Comeback,'" *New York Times*, August 11, 1932.

12. "Igloo, Byrd's Dog, Polar Hero, Is Dead," *New York Times*, April 22, 1931.

13. "Igloo, Byrd's Dog, Buried," *New York Times*, June 1, 1931.

14. E. A. Torriero, "Kabul Zoo's Shell-Shocked Lion Dies," *Chicago Tribune*, January 27, 2002, http://www.chicago.tribune.com.

15. "*Denver Post, Rocky Mountain News* to Charge for Obituary Notices," *Denver Post*, March 24, 2002, EBSCO Host online (accessed October 27, 2003).

16. Cara DeGette, "Public Eye: Free Obituaries; A Dying Breed," *Colorado Springs Independent*, April 10, 2002, http://apw.softlineweb.com/printWin.asp?recNum=2&recformat=articlecitation&scope=current (accessed June 10, 2003).

17. Michael Roberts, "Dead Lines: The Denver Dailies Change the Way They Handle Obituaries—For Better and for Worse," Westword, April 18, 2002, http://apw.softlineweb.com (accessed June 10, 2003).

18. Paraphrase of Tom Peterson, in Tom Peterson, "Media: Your Fame, Their Fortune; C-J Moves to Paid Obituaries," in *LEO: Louisville Eccentric Observer*, July 25, 2001, http://apw.softlineweb.com (accessed June 10, 2003).

19. Peterson, "Media."

20. As of early 2000–2001, the *Sun* did run pet obits with human ones as a matter of policy, but I am not aware whether this is still the case. At that time they had run only four of them that year,

but intended to continue. Their inauguration of this policy was right around the time that the contestations erupted nationally.

21. Debra Mckinney, "Newspaper Obituaries Pay Tribute to Pets," *Anchorage (Alaska) Daily News*, April 7, 2002, http://www.newsbank.com/nl-search/we/Archives?p_action=doc&p (accessed October 27, 2003).

22. Jenny Walker, retired obit wrier for the *Durham (N.C.) Morning Herald*, communicated this story to me at the Obituary Writers conference, and provided the details in a follow-up e-mail on June 20, 2008, but could not remember the exact name or the date of the obituary. Interestingly, her mother had also improvised a way to publicize an animal's death. She once took out a classified ad in the local Chapel Hill, North Carolina, newspaper "announcing (the dog's) death to all of his friends" in town. This positions the people of the local community as mourners, assumes they have both a desire and right to know, and manipulates the categories of the newspaper to achieve an otherwise prohibited goal of providing a dog's obituary for the public record. This act of resistance or creative improvisation is also seen in pet cemeteries and in some pet burial practices, where, for example, a pet is snuck into a casket of a person who died at the same time (Michael Lensing, Lensing Funeral Home, Iowa City, Iowa, personal communication, 2006).

23. *Tacoma (Wash.) News Tribune*, April 25, 2001.

24. Joann Loviglio, "*Philadelphia Daily News* Adding Pet Death Notices," Associated Press, March 5, 2002, http://web.lexis-nexis.com (accessed October 27, 2003).

25. Kathryn S. Wenner, "Dog Gone, Not Forgotten," *American Journalism Review* 24, no. 4 (May 2002): 14–15.

26. Bob Levey, "Latest in Questionable Taste: Pet Obituaries," *Washington Post*, May 1, 2002, http://web.lexis-nexis.com (accessed September 25, 2003).

27. Lewis Grossberger, "Spot Dead! City Mourns," *MediaWeek*, June 5, 2002, 38, accessed through EBSCO database on June 11, 2003.

28. Levey, "Latest in Questionable Taste."

29. Levey, "Latest in Questionable Taste."

30. Similar comparisons of value emerge in the CNN segment devoted to discussing pet obits. See the transcript for CNN Talkback Live, airing on March 6, 2002, transcript #030600CN.V14, retrieved through LexisNexis on October 3, 2003.

31. For two research-based discussions of this lack of support and the mourning and burial rituals that people develop to combat it, see Tami Lynne Harbolt, "Too Loved to Be Forgotten: Pet Loss and Ritual Bereavement" (master's thesis, Western Kentucky University, 1993); and Elizabeth Van Loo, "A Study of Pet Loss and its Relationship to Human Loss and the Valuation of Animals" (Ph.D. diss., Troy State University, 1996). Loo writes: "often their [pet lovers] grief goes unrecognized and proper support for their bereavement is unwittingly denied them by society. As a result, their grief becomes disenfranchised" (12).

32. I explore these issues of online mourning further in "Relating to Animals after Death: Virtual Pet Cemeteries and the Public Act of Grieving Online," paper presented at the 2003 meeting of the International Anthrozoology Association conference at Kent State University, in Canton, Ohio. Numerous contemporary popular press books exist on coping with pet loss, especially those written by psychologists and counselors for the lay public. Among the best academic articles examining historical practices of mourning for pets and the representation of such is Teresa Mangum, "Animal Angst: Victorians Memorialize Their Pets," in *Victorian Animal Dreams: Representations of Animals in Victorian Literature*, ed. Deborah Denenholz Morse and Martin A. Danahay (Hampshire, England: Ashgate, 2007), 15–34.

33. Natalie Pompilio, "The Brides Wore White," *American Journalism Review* (October 2002): 12. By the time the *New York Times* made this announcement, 36 of the top 100 papers in the United States already accepted such announcements, at least in theory, reports Carl Sullivan, "Same-Sex Unions Gain Notice; But Papers Not Wedded to Policies," *Editor and Publisher Magazine*, September 2, 2002, 6.

34. I thank Jean Olsen of Santa Barbara, California, for sending me this clipping.

35. Marylin Johnson, *The Dead Beat: Lost Souls, Lucky Stiffs, and the Perverse Pleasures of Obituaries* (New York: Harper Perennial, 2007), 31.

36. Johnson, *Dead Beat*, 35.

37. Anon., *Honoring Your Loved One in Print: Your Guide for Writing and Submitting a Personalized Obituary"* (Champaign, Ill.: *News-Gazette*, 2003), 3.

38. Anon., "Honoring Your Loved One in Print," 7.

Golden Retrievers Are White, Pit Bulls Are Black, and Chihuahuas Are Hispanic

Representations of Breeds of Dog and Issues of Race in Popular Culture

MEISHA ROSENBERG

As the buzz surrounding the Obamas' recently acquired Portuguese water dog shows, choices about dog breed carry political implications having to do with identity and race. Initially, President Obama expressed interest in adopting a shelter or rescue dog—"Obviously, a lot of shelter dogs are mutts like me," he said, using a little anthropomorphic humor to make reference to his own mixed-race heritage.[1] But partly because the family's other major concern was daughter Malia's allergies, he ultimately accepted a pedigreed Portuguese water dog, Bo. Bo had been originally bred in Texas; when he did not work out with his first family, he went to the trainer of Senator Edward Kennedy (Kennedy owned another dog from the same litter). Senator Kennedy then gave Bo as a gift to the Obamas. Bo, while he was not acquired directly from a breeder, also was not a shelter dog or a mongrel. His identity was in a gray area that upset some advocates who had wanted the president to stick by his initial promise and adopt a shelter dog. Obama's acceptance of Bo signaled acceptance of Kennedy's Democratic mantle, at the same time that it demonstrated almost pitch-perfect bipartisanship in terms of dog politics.

The Humane Society of the United States (HSUS) held a campaign in which one could send a thank-you card to the president for stating his interest in a shelter or rescue dog. The HSUS also congratulated the first family on adopting a second-chance dog, one who had been returned by its first owners.[2] However, there was some disappointment in the animal welfare community, as Wayne Pacelle, chief executive of the HSUS, acknowledged: "He's in a gray area," Pacelle said of Bo. "But I will say that many animal advocates are disappointed that he (Obama) didn't go to a shelter or breed rescue group, partly because he set that expectation and because so many activists are focused on trying to reduce the number of animals euthanized at shelters."[3] The HSUS employed the liberal rhetoric of saving the innocent strays; the American Kennel Club (AKC) argued the conservative party line of allegiance to pedigree and breed. The AKC, after running an online campaign in which people could vote online for their breed pick, stated they were pleased that the Obamas would be modeling responsible purebred dog ownership. Obama walked a tightrope between the two camps.

What has not been mentioned about Obama's decision so far is the way in which race played into both points of view. His choice of Bo reads, in this sense, as an acceptance of pedigree and all it implies. At the same time, the dog's status as a "quasi-rescue," in the words of Pacelle, suggests modified allegiance to programs of rehabilitation and reconciliation for society's others. The figure of Bo served to cloak as much as to reveal race politics for this presumably "postracial" presidency.

In this chapter, I show how the trope of pedigree dog breeding and, with it, representations of specific breeds symbolize race and class hierarchies in the media. While a dog may be used in advertising, film, or animation to symbolize any number of different qualities (Nancy Spears and colleagues identify "mount, messenger, companion, guardian," and "hound of hell," among the associations for dogs[4]), status and racial identity should be considered among them.

Other scholars have identified class as important in cultural constructions of pets and dogs in different time periods (in particular, Thorstein Veblen in his *Theory of the Leisure Class*; James Serpell in his *In the Company of Animals*; and Harriet Ritvo in her *Animal Estate*).[5] Issues of race as linked to dogs and dog breeding are also mentioned by Ritvo and discussed by Mark Derr, who writes that by the mid-1880s, "Belief in the superiority of the pure-blood dog and of scientific breeding were of a piece with the justifications of class, racism, and eugenics."[6] Discussing contemporary issues, Jonah Goldberg argues point blank in two online articles that "Westminster is racist" (Westminster is the benched dog show held every year in Madison Square Garden that produces a number of best of breeds and a "best in show"). He points out that eugenics—the late-nineteenth-to-early-twentieth-century belief in improvement through enforced sterilization, applied prejudicially to people with disabilities, those in the lower-class, and nonwhites—persists in dog showing and the emphasis on pedigree. In addition, *Pedigree Exposed*, a British Broadcasting Company (BBC) investigative television documentary aired in 2008, criticized dog breeders for the health issues endemic to inbred dogs and highlighted the connection to eugenics.[7]

These examples point out the ideological connections between the activity of pedigree dog breeding and nineteenth-century theorizing about race in England and America. I argue that these connections inform contemporary public perceptions of dogs in terms of race. Before I discuss contemporary examples, however, it is important to review historic conceptualizations of the dog pedigree.

PEDIGREE AND PURITY IN THE VICTORIAN ERA

Representational certainty and familial fidelity are the presumed guarantees of the pedigree. We look at a dachshund and we assume we know what we are looking at because it is a dachshund. That the dachshund may not know or care about this is part of the equation of human control. Judges at Crufts and Westminster do not look at how well a dachshund might ground a badger (the traditional activity associated with this breed); they evaluate its theoretical capability to perform such tasks through appearance. Conformation, or adherence to physical standards, is how a purebred dog is measured. This system is a relatively new invention.

Some breeds, such as the Pekingese and the greyhound, have been bred for centuries. However, in England, previous to the creation of the Kennel Club in 1873, dogs had mainly been classed by the jobs they performed. They were known as hunting dogs, guard dogs, or house dogs. Harriet Ritvo writes, "The very notion of breed as it was understood by Victorian dog fanciers—a subspecies of race with definable physical characteristics that would reliably reproduce itself if its members were crossed only with each other—may have been of relatively recent origin."[8] The showing of pedigreed dogs divorced the dog from its working history and behavior, favoring a symbolic over a functional appreciation. Additionally, the men who founded the British Kennel Club and the AKC (the AKC started in 1884) were concerned with distancing themselves from the lower-class activities of public-house dogfighting and rat-killing contests. Championship promised entry into a powerful world of MPs, duchesses, and earls.

New breeds of dog were developed (the clumber spaniel, the golden retriever), and old designations were subjected to scrutiny at the same time that pseudo-scientific systems like phrenology quantified human physical traits to fit race and class hierarchies. The Kennel Club came into being at a time when English society was preoccupied with locating provable dimensions for national, ethnic, and racial differences. Evolutionary theory caused many people to reexamine the relationship between humans and animals. Ritvo states: "The institutions that defined the dog fancy projected an obsessively detailed vision of a stratified order which sorted animals, and, by implication, people into snug and appropriate niches."[9] The collection of exotic specimens of plants, animals, and people from far-flung colonial outposts and their exhibition as curiosities in events such as the World's Columbian Exposition were activities arising from the same cultural matrix that fostered the development of the kennel clubs.

Metaphorical linkages of breeds of dog with human class and race categorizations are common in the writings of Victorian-era breeders and fanciers. In *The Practical Kennel Guide*, Gordon Stables wrote: "It is, too, a strange fact, that the more highly civilised a nation is, the greater is its care and culture of the canine race, and the more highly bred are its dogs. Look at China, for example, or even native India—whose semi-civilisation seems to have been crystallised in the bud many, many hundreds of years ago—look at these nations, and look at their dogs—mongrel, gaunt, and thievish, and only half reclaimed from the wild state."[10] According to Ritvo, "the dichotomy between domesticated animals and wild animals was frequently compared to that between civilized and savage human societies." She relates that Charles Darwin "cited a report that two Scotch collies who visited Siberia 'soon took the same superior standing' with regard to the native dogs 'as the European claims for himself in relation to the savage.'"[11]

Well into the mid-twentieth century, racist ideologies were fueled by pseudo-scientific analyses of physiognomy. Harvard physical anthropologist Earnest Hooten and others posited that criminals and nonwhites existed closer on the evolutionary chain to the animal world.[12] The image of an ape was repeatedly used to depict Africans and other groups of people perceived to be lower than whites. Lewis Perry Curtis points out that in the late nineteenth century, "Virtually every country in Europe had its equivalent of 'white Negroes' and simianized men, whether or not they happened to be stereotypes of criminals, assassins, political radicals, revolutionaries, Slavs, gypsies, Jews, or peasants."[13]

Like images of the ape, representations of the dog also played a role in the theorizing of race as a physical category. In 1895, Sir Everett Millais argued that inbreeding is a natural phenomenon that gives rise to discrete "specimens":

The domestic animals and the white man—that is, the higher types of man—are of newer creation or evolution than the wild or dark; consequently their type is less fixed than the older ones, and when they by any circumstances are bred together the newer type goes down before the old. Now, gentlemen, I need hardly say to you that if a man wishes to succeed as a breeder, or, rather, as a cross-breeder, with views such as I entertained when crossing the Basset with the Bloodhound, he must, to a certain extent, be conversant with these questions.[14]

As further evidence, he reports that in South America, "If you take the half-caste and mate her with the white man, it takes some five generations before you can breed out the negro's blood."[15] Millais's statements reflect a belief during this time period about white purity and the contaminating effects of the blood of nonwhites. Millais goes on to explain that, while inbreeding produced pedigree, he nonetheless had to prevent "degeneration" in his basset hounds through outcrossing. His interweaving of descriptions of dog breeding and human mating shows how conceptions about the two were closely linked at the time.

As Mark Derr and many others have noted, in America, "From the start, Indians and wolves were discussed in much the same language, as wild, brutal, savage, uncivilized creatures blocking the advance of Christian civilization."[16] These earlier equations of wild or native dogs and wolves with Native peoples set the stage for a long-standing analogy aligning dogs with the domestication or eradication of the other. Victorian breeders did not invent the thinking equating human "savage" to wild dog, and civilized human to well-bred dog, but they applied these analogies about animals to their pseudo-scientific understanding of race.

Through representations of dogs, humans now say things about race that would otherwise go unsaid. Through dogs, popular culture speaks about its assumptions and its anxieties having to do with difference. At the bottom of many of these anxieties are the questions: What makes us human? What or who is domesticated, and what or who is wild? Portrayals of dogs work these concerns out through the category of breed, tapping into assumptions about pedigree, heredity, and race.

DOGS, BREED, AND RACE NOW

Social scientists and historians define race as a cultural category or construct that identifies groups of people with relation to national and ethnic heritage. In other words, race is not quantifiable; each human being has many thousands of ancestors, and we share common origins. Variability and group mixing mean that human populations are in constant flux. Since the civil rights era in the mid-twentieth century, Americans' references to race are frequently coded and constitute what, in *The Race Card*, Tali Mendelberg calls "implicit appeals," writing, "The enormous shift in public norms of racial discourse . . . [after the mid-twentieth century] created a near-universal tendency to self-censor. Some people censor themselves because they aspire to be egalitarian, others because they wish to conform to the social pressure of norms of discourse."[17] Eduardo Bonilla-Silva identifies what he calls the discourse of "color-blind racism" among college students and adults he interviewed, in which projection, diminutives (minimizing one's beliefs), and an extension of "cultural rather than biological explanation of minorities' inferior standing" characterize race talk.[18] When admission of racial views has become highly coded and hidden, the dog is a convenient wrapper for subtle racial messages and associations. Such messages may influence perceptions of groups of people; they may also impact views of and treatment of dogs, whose realities become secondary.

Preoccupations with racial and class hierarchies through pedigree have been transformed by American consumer appetites into a dizzying array of niche markets. Walk into any PetCo or PetSmart, the chain pet-supply stores, to witness the emphasis on a dog's breed heredity in everything from food to magazines. Pedigree, a major American dog food company, alerts us to the importance of pure ancestry in its very name, while a slew of glossy calendars and magazines are all predicated on breed identification (at my local mall I can find calendars of golden retrievers, dachshunds, German shepherds, and beagles). A dog is not just a dog—and one is not so much a "dog lover" as one is encouraged by this market to be a diehard golden retriever enthusiast or a border collie lover. *Dog and Kennel Magazine* caters to owners of specific breeds (each month features a different model specimen), while each breed has its own universe of publications (such as *Doberman Digest* and the "Long and the Short of It," a blog about dachshunds).[19] There are also organizations proclaiming that "mutts," meaning mongrels, are the best, such as a rescue organization called Mixed Up Mutts.[20]

In this diversified field, the dog acts as an extension of the self in a way that can figure race. Michelle, an interview subject in a *Journal of Business Research* article by Jill Mosteller, discusses the rat terriers she breeds, and why she prefers them to other types of dogs, in terms of self-identification and culture: "Dogs are . . . like people, some people are more cultured than others."[21] Mosteller's example shows how dog enthusiasts sometimes speak in code through their dogs about their own preferences and biases having to do with heredity, race, and cultural traits.

Dogs may serve as particularly good vehicles for race talk today because they inhabit a particular place of cultural unease about the nature of identity in both humans and dogs. The dog genome has been mapped; a company called the Canine Heritage Breed Test promises to "unravel the mystery of your mixed-breed pup"; dogs have been cloned (an Afghan hound in South Korea; a mongrel, Missy, and a yellow lab in the United States).[22] The control of dog appearance and identity can happen at the level of DNA.

In the age of the genome, a purebred dog can signal the possibility of inherited "purity" as well as the specter of its opposite—the threat of contamination from the mongrel or unpedigreed dog. In an excellent essay on the Missyplicity Project (in which Missy, a pet mongrel dog, was cloned), Susan McHugh explains that "breeding records, as the necessary evidence of human control required of all registered members of canine breeds, legitimate not only an interspecific (or cross-species) sphere of intimacy but also a hierarchy projected by the human onto canine social spheres. This hierarchy primarily serves to distinguish the breed dog from the mongrel, whose lack of documented parental matching confirms her inability to meet the otherwise visual 'qualifications' of a 'blood-dog.'"[23]

In other words, cloning and pedigree remedy, through representational fidelity, the disorder of the mongrel's body and animal sexuality. In McHugh's reading, the rhetorical goal of the Missyplicity Project is to erase intra- and interspecies disorder through a textual and genetic coding of Missy's identity. McHugh points out the centrality of representation: the difference between a pedigreed and a nonpedigreed dog is quite literally that the pedigreed dog is brought into existence because it is representable by humans in textual or visual form. While mongrels are of course also represented in advertising, novels, and film, they are often portrayed in ways similar to Missy: narratives erase mongrel history through anthropomorphic resolution (see the later discussion of *Beverly Hills Chihuahua*). Mongrels (and pit bulls, considered an unofficial designation) are equated with a human other and either demonized or rehabilitated (sometimes both). Specific dog breeds or types are often represented in ways that correlate them to human racial categories.

Next, I look at a series of ads featuring golden retrievers as representations of white middle-class family life; at pit bulls as symbols of lower-class African Americans; and as Chihuahuas as stereotypes of Latinos. I have chosen these three because they most obviously illustrate some of the issues at hand and because they happen to be currently popular breeds seen in a variety of formats. I do not at all intend to suggest that anyone who has or breeds one of these types of dogs is a racist, or that representing qualities through dog breeds automatically means one is prejudiced. There are certainly many examples of images of dog breeds used to advertise or convey any number of qualities not related to race, and other reasons why advertisers might choose a particular animal (for example, trainability or size). What I mean to say, rather, is that the history of the dog-human relationship has been marked by racial concepts and that advertisers, celebrities, and media representatives, when they portray dogs, tap into those concepts, sometimes knowingly and sometimes not.

AMERICA'S GOLDEN BOY

Golden retrievers and yellow Labrador retrievers are extremely popular in contemporary media and advertising as accessories to white upper-middle-class life. America's favorite breeds in 2009 according to AKC registration statistics are the Labrador retriever followed by the German shepherd, Yorkshire terrier, and golden retriever. These breeds arose from Anglo-Saxon hunting culture: according to the AKC, "all retriever breeds begin with the water-loving Saint John's dog of Newfoundland," and the golden retriever originated from Lord Tweedmouth's estate in the Highlands of Scotland, while the Labrador was so-named by an English earl.[24] In contemporary representations, these light-colored retrievers are figured as the popular dog of white nuclear families.

In a magazine advertisement for metal roofing, the only face in the ad is that of a golden retriever looking at a white female, who sits in a lawn chair next to a man. The ad reads: "4 kids. 11 grandchildren. 9 cars. 2 dogs. 3 remodels. 66 vacations. 1 roof." The race and class markers are clear: the couple is white; the man is wearing a polo shirt and shorts; the house is white and has an expanse of green lawn.

Another ad with a golden retriever portrays white upper-middle-class life. This one shows General Electric frontload laundry machines in the background, with a blond woman and her blond children in the foreground with their golden retriever puppy. The mother is helping the children with their soccer shoes. The advertisement reads: "Until now, an attractive, fully-featured frontload laundry pair was a luxury for the privileged few. Not anymore." The message is the same as that for the metal roof: wealth and the upper-middle-class life are affordable; and the markers to it are clear and include having a golden retriever and being, or at least acting, white. Whiteness is apparent not only in the blondness of the family and their dog but also in the white-colored machines, the white background of the ad, and the activity of cleaning suggested by the machines.

In a series of L. L. Bean catalog ads, golden retrievers symbolize white American masculinity and focus on the athletic body. One dog is in a pickup truck behind a smiling white man; another golden retriever jumps into a lake with a group of white models, embodying physicality, recreation in an Adirondack-like setting, and echoes of the historical setting of the retriever's provenance.

In an H&M ad, a group of children, mostly white, with the exception of a dark-skinned boy in the center, are accompanied by a seated golden retriever. The lightest child in the ad, a blond girl, is visually central as she looks at the camera and has her arm in a gesture of ownership and belonging around the golden retriever. The retriever is symbolically equated with the blond girl.

In a series of television advertisements for Cottonelle, the toilet paper, we meet a yellow Lab puppy (whose fur color is more white than yellow). He speaks in a male voice to testify that using the toilet paper will put any consumer in "the lap of luxury." We see a white woman in a luxurious apartment at the beginning of one commercial. The puppy is compared directly to sheets of toilet paper itself as the female voice says, "A roll of Cottonelle toilet paper is so much longer than a roll of the leading brand's Red Pack, that if we had a puppy for every extra sheet." "There'd be a hundred of us," the puppy responds, as it appears with dozens of identical puppies.[25] This invokes the image of dog clones achieved through breeding or genetic manipulation. The Cottonelle puppy represents both the animal body as site of the eliminatory function and the control of the animal body through cloning. The white color of the dog, the toilet paper, and the white woman, similar to the General Electric ad, suggest that whiteness is a quality dogs can simultaneously display and hide through symbolic absorption.

The popularity of the yellow Lab and golden retriever as images of mainstream values extends to the world of books and popular movies. The mega–best seller *Marley and Me* (followed by the 2009 movie) starred a yellow Labrador retriever as the goofy, lovable family pet. The twentieth edition of the official AKC guide, *The Complete Dog Book* (2006), shows a golden retriever family on the cover (an adult dog with two puppies resting on her back). Other recent books about dogs featuring yellow Labs or golden retrievers on the cover include Mark Doty's *Dog Years*, Elizabeth Marshall Thomas's *The Social Lives of Dogs*, and Jon Katz's *The New Work of Dogs*.[26] All these popular books—*The Social Lives of Dogs* and *Dog Years* were both *New York Times* best sellers—signal dogs' elevated status as objects of literary concern and upper-middle-class leisure.

Precisely because the public can be reliably counted on to associate golden retrievers and yellow Labs with a constellation of values figured as upper-middle class, their coding as racially white can go unnoticed. It is not any one particular image or commercial that states, more than any other, the association of golden retrievers or yellow Labs with whiteness—it is rather the ubiquity of these images. In a culture where explicit racial images and language are taboo, golden retrievers and yellow Labs can safely transmit assumptions about whiteness.

PIT BULLS

Golden retrievers' signification with markers of white identity is usually unremarked, in the same way that whiteness itself is usually unremarked: it is conceived of as a neutral signal for cultural norms. It is harder to miss the media's symbolism of racialized otherness in the figure of the pit bull. Pit bulls generally appear in the media in two guises: as misunderstood worthies saved from the brink of destruction (many celebrities and writers have adopted them) or as dangerous dogs. They do not appear, as golden retrievers do, as the common family dog, although during the 1950s they did. One might point out that the bull terrier used in Target advertising and Spuds MacKenzie, the Budweiser bull terrier, are in the same family as pit bulls; however, they appear more in the specific context of the brand mascot, and their unusual bullet-shaped nose-bridge makes them visually distinct from pit bulls.[27]

Usually media portrayals of pit bulls are negative, and they frequently link the dogs with black men. For an example of this, one need look no farther than the Michael Vick case. As Wright Thompson wrote for ESPN regarding the case, "Animal rights activists think it's about cruelty. . . . African-Americans in Atlanta, according to prominent black leaders, think it's about Vick not getting due process because of the color of his skin."[28] The media portrayed the gruesome realities of dog fighting, while People for the Ethical Treatment of Animals (PETA) protests made it seem as if Vick was already guilty before his trial. The outcry against dog fighting was intense—but did the nature of the crime warrant the attention? Regardless of one's point of view, the Vick case highlighted the symbolic connections mass media draws between black men and pit bulls. Playing on these same linkages are shows such as *Animal Cops*, the weekly reality show on the Animal Planet network filmed, among other places, in New York, Detroit, and Philadelphia. One episode, "Extreme Danger," portrayed "attack" Rottweilers owned by an African American man. Such programs often sensationalize animal neglect and abuse in the context of cities with low-income populations of African Americans.

Where mainstream media often demonizes pit bulls, some African American artists have appropriated them as powerful symbols of identification. Long a part of the public imagery of pit bulls has been Snoop Dogg, who himself has pit bulls (on one episode of *Fatherhood*, his reality show, he claims he and his family have twenty dogs that they breed). In his music he proudly claims his identity in canine terms that align him with a symbolic underdog. He uses the term "dog" to signify toughness, loyalty, and a criminal past. He does so within an African American tradition of using the word and the concept to refer to a complex mix of elements of identity. With Snoop Dogg, one never knows if he's being completely serious. Snoop Dogg's album titles all feature the word "dog," some humorously: *Doggfather*, *Doggystyle*, and *No Limit Top Dogg*. Other rappers have adopted the moniker in various forms (Nate Dogg is another rapper and Snoop Dogg's cousin; Tha Dogg Pound is another rap group that signed with Death Row Records, a label with which Snoop Dogg has recorded. Lil' Bow Wow's albums have included *Unleashed* and *Beware of Dog*.)

One of Snoop Dogg's most famous songs, from 1993, "Who Am I (What's My Name)?" samples from George Clinton's song "Atomic Dog." In the lyrics, "dogg" conjures up a whole chain of associations from the word in African American culture: the "woof" refrain, sampled from George Clinton's "Atomic Dog," from the problack funk of the 1970s, is like the black fraternity Omega Psi Phi's "woof" call (used like a secret handshake). Pledges of Omega Psi Phi are referred to as "Que-Dogs." Arsenio Hall also adopted this "woof" call on his shows, where audience members sat in the "dog pound" near the stage, "woofed," and pumped their fists in the air. DoVeanna S. Fulton recognizes this reaction as an essentially positive, problack one: "This particular form of applause has long been a part of African American youth culture."[29]

Many associations are attached to Snoop Dogg's use of the symbol of the dog, and they include group and family belonging, sexual prowess, the idea of the underdog, and the power gained from an animal mascot or emblem. His use of *Doggystyle* as one of his album titles makes the sexual identification clear; so do his lyrics' frequent use of the word "bitch" to describe women. Dogs are also family in his symbolic universe, and he is often portrayed as being affectionate toward his pit bulls on *Fatherhood*. Snoop Dogg's identifications with pit bulls also tap into the symbolic resonances around dogs and African American history. His sometimes self-satirizing dog-identified stance speaks to a past in which rhetoric often equated nonwhites with sexualized animals. Snoop's "dogg," instead of being an animal historically used by whites to maintain power through defending territory, killing interlopers, and tracking escaped slaves, becomes both a reenactment and a reversal of that legacy.

Another rapper, DMX, like Snoop Dogg, uses dogs as symbols of oppression, but his videos are more explicitly violent. His lyrics also identify him with dogs and underdogs. His video for "What's My Name" (on the album *And Then There Was X*, 1999) caused controversy because it depicted pit bulls in a fighting ring.[30] In the video, the dogs are on chains, and DMX is wearing a similar thick chain. DMX stands alone in the center of the ring and says "Bust my gun / . . . I'm always down for the one on one / Dog for life." The lyrics identify both DMX and his imaginary opponent as symbolic dogs. While the chains on the dogs reflect the culture of dog fighting, they also refer to the shackles of slavery. In the video, two men hold two dogs who growl, posture, and lunge. For DMX, "dog" fulfills the embattled stance of a lone man who insists on fighting "one on one" and stands outside the scope of the law. In 1999, thirteen neglected pit bulls were found at his home, and in 2002 DMX pleaded guilty to charges of animal cruelty. While his case was not sensationalized to the degree of Vick's, it was another example of the associations the media makes between pit bulls and black men. Both DMX and Snoop Dogg appropriate these associations in their music and videos, sometimes satirically (as with Snoop Dogg) and sometimes seriously.

The associational linkages set up by various media between pit bulls and black men extend beyond the symbolic when they influence behavior and legislation. Malcolm Gladwell, in *The New Yorker*, drew a parallel between pit bull bans and racial profiling by the police, making the point that by banning pit bulls, municipalities focus on breed, not other factors that might be more important (such as whether dogs are on or off leash at the time of the bite incident; or whether the animal in question has been spayed or neutered).[31] That jurisdictions continue to focus on the dog's breed enforces a kind of anthropocentric racialism in legislative approaches to animals. Humane shelters and sites like Petfinder.com shy away from using the term "pit bull," fearing that the public will not want to adopt animals bearing so many negative associations (instead of using "pit bull," they often use "terrier").

Similar to the avoidance of pit bulls, potential adopters, especially in the South, pass by black dogs, preferring their lighter compatriots in what is known in shelters as "Black Dog Syndrome." Some speculate that black does not show up well in online photographs;[32] perhaps it also has something to do with the many folk legends associating bad luck with black animals. Web sites such as blackpearldog.com couch the problem in terms of a minority group that has been victimized through no fault of their own. Whether or not one describes the phenomenon as a projection of racism, it seems clear that coat color in pet dogs can have great meaning for humans.[33] The marginalization and now rehabilitation of black dogs do suggest that when it comes to assigning value to pets, humans carry a lot of baggage related to color, race, nation, and class.

CHIHUAHUAS IN HOLLYWOOD

Representations that link dogs with racial groups can be promotional, as in the golden retriever ads and Snoop Dogg's identifications with dogs, or they can transmit negative associations, as in the media demonization of the pit bull and black men. Because some dogs are seen as cute and harmless, especially in animated and children's movies, they can transmit damaging ideas about racial identity under the guise of fun and games.

One example of such a linkage occurred in the late 1990s when Taco Bell had a series of commercials with a Chihuahua who said "*Yo quiero* Taco Bell" (I want Taco Bell). The ads were extremely popular, spawning T-shirts, and the phrase made its way into the vernacular. However, some Latinos took offense. In *El Tecolote*, Darren Villegas said that the Chihuahua in the commercial looks like a rat, "So it looks more like a pest or a nuisance, which is probably how some upper middle-class white people may subconsciously relate to Latinos."[34] In addition, some of the ads featured the Chihuahua wearing the characteristic black beret worn by Che Guevara and using the rhetoric of populist revolt: "So now that we know what the people think," says the 1950s-era style voiceover, with Latin music playing in the background, "let's hear what our glorious leader has to say." Then we hear the "voice" of the Chihuahua saying, "*Hasta la vista*, Whopper." In this example, a breed of dog historically from Mexico is used to represent, in advertising, the Argentinean revolutionary Che Guevara. While some consider this humorous satire, others are offended, both by the trivializing representation of a man some consider to be a killer and, at another level, by the equation between a small, ratlike dog and a group of people.

In the 2008 Disney film *Beverly Hills Chihuahua*, once again the Chihuahua breed represented Latinos in a trivializing and stereotyped way. Chloë, a female cream-colored Chihuahua who is the main character (voiced by Drew Barrymore), leads a life of privilege, wearing coordinated outfits with jewelry. She regularly goes to the groomer to gossip with her dog friends and attends lavish dog birthday parties (photographed for the celebrity pup magazine "*Purebred Monthly*") with her owner, Vivian (played by Jamie Lee Curtis). Chloë says about herself, "I was born to shop, not fight, unless it's a sale." Chloë is looking for a certain kind of dog of the opposite gender: "It's not easy to find a mate with papers let alone one you can love." This double play on "papers" is especially offensive, as it refers to both a dog's pedigree and to an immigrant's Green Card or naturalization certificate. But it is further evidence that in American popular culture, pedigree equals white privilege.

Further plot developments show that Disney uses the dog characters to develop a theme of race and class. Chloë meets Papi, the Chihuahua owned by the landscaper. Papi (voiced by George Lopez ["I put the 'Wow' in Chihuahua"]) does not initially meet her criteria—he speaks with a Spanish accent, and calls her "*Mi Corazon*." Chloë's figuration as a white upper-class dog is contrasted to Papi's identity as a lower-class and darker-haired Chihuahua (the dog playing his role was described in a *People* magazine article as a Chihuahua mix from a shelter).[35] When Vivian leaves on a trip and her niece is tapped to dog-sit, all hell breaks loose: the irresponsible niece takes the dog to Mexico, and Chloë gets lost. She is trapped by someone running a dog-fighting ring, and the street dogs who are her cell mates wonder why she does not speak Spanish. "Why would I speak Spanish?" she asks. "Hello, you're a Chihuahua *Miha!*" a large brown mixed-breed dog replies in a Spanish accent, underlining the stereotype that all "true" Latinos must speak Spanish.

In Mexico, Chloë stumbles upon an ancient empire of Chihuahuas in a crumbling Aztec city led by a long-haired Chihuahua named Montezuma, who tells her, "The Aztec people left long ago, but we remain." Here he and his Chihuahua army create a fantasy of aboriginal identity. The promotional image for the movie shows the hero and heroine standing before a ceremonial scene in front of a pyramid, with the Chihuahua army seated around them. It is the city of Chihuahuas who ultimately get Chloë to learn to stand up for herself. Chloë's dignity and ethnic pride are restored, as is her openness to Papi's advances when they reunite in the end. The idea that her upper-class life meant she was out of touch with her Mexican heritage drives the story. By the same token, Papi's upwardly mobile arc happens through association with the upper-class white-colored Chihuahua. (*Lady and the Tramp*, a Disney production about a pampered female spaniel and a street mongrel, follows a similar arc.) Racial reconciliation occurs through the anthropomorphic mating of two Chihuahuas.

The likening of a dog or type of dog to a person or group of people is not a new phenomenon. And certainly dogs are not alone in the genre of funny animal stories: Bugs Bunny was voiced by Mel Blanc and introduced the world to a wisecracking Jewish New York style; the 2007 movie *Ratatouille* featured a French rat who became a gourmet chef. But generally, in movies and ads, bunnies are bunnies and rats are rats—the metaphor travels one way only, to assert human traits where none exist. (No one gets a pet bunny assuming it will confer Jewish New Yorker attitudes or status—although *Finding Nemo* seemed to inspire the increased purchase of clownfish in 2003.)[36] But repeated representations of dogs do affect perceptions of both dogs and people. Concepts about race that formed in tandem with the development of the kennel clubs continue to inform portrayals of dog breeds. It is important to acknowledge the many associations we carry about the breed identity of dogs and our racial or class prejudices, not only because they reveal much about mainstream attitudes but also because these assumptions become part of our lives with dogs.

NOTES

1. Anon., "Obama's First Press Conference Turns to Dog Talk: 'This Is a Major Issue,'" Associated Press/Huffington Post, November 7, 2008, http://www.huffingtonpost.com/2008/11/07/obamas-first-press-confer_n_142201.html.
2. Humane Society of the United States (HSUS), "Online Card Thanks Future First Family for Choosing to Adopt a Dog," press release, November 25, 2008, http://www.hsus.org/press_and_publications/press_releases/online_card_thanks_obamas_for_choosing_adoption_112508.html; HSUS, "Second Chance Dog Heads to the White House," press release, April 12, 2008, http://www.hsus.org/pets/pets_related_news_and_events/second-chance_dog_heads_to.html.
3. Sharon Theimer, "Promises, Promises: Is Obama Dog a Rescue or Not?" Associated Press/Breitbart.com, April 13, 2009, http://www.breitbart.com/article.php?id=D97HQQP00&show_article=1.
4. Nancy E. Spears, John C. Mowen, and Goutam Chakraborty, "Symbolic Role of Animals in Print Advertising: Content Analysis and Conceptual Development," *Journal of Business Research* 37, no. 2 (1996): 87–95.
5. Among those who have written about class as a factor in pet ownership are Thorstein Veblen, *Theory of the Leisure Class: An Economic Study in the Evolution of Institutions* (1899; repr., New York: Penguin, 1979), 45; James Serpell, *In the Company of Animals: A Study of Human-Animal Relationships* (New York: Basil Blackwell, 1986), 34–47; and Harriet Ritvo, *The Animal Estate: The English and Other Creatures in the Victorian Age* (Cambridge, Mass.: Harvard University Press, 1987).
6. Mark Derr, *A Dog's History of America: How Our Best Friend Explored, Conquered, and Settled a Continent* (New York: North Point Press, 2004), 238.

7. More sensationally, PETA held a protest in which members dressed in KKK outfits outside Westminster in 2009. For more on this, see Gene Farris, "PETA Dresses in KKK Garb Outside Westminster Dog Show," *USA Today*, February 10, 2009, http://www.usatoday.com/sports/2009–02–09-peta-westminster-kkk-protest_N.htm.

8. Ritvo, *Animal Estate*, 93.

9. Ritvo, *Animal Estate*, 93.

10. Gordon Stables, *The Practical Kennel Guide, with Plain Instructions How to Rear and Breed Dogs for Pleasure, Show, and Profit* (London: Cassell Petter & Galpin, 1877), 13.

11. Ritvo, *Animal Estate*, 16.

12. The Eugenics Archive, Dolan DNA Learning Center, Cold Spring Harbor Laboratory, thematic summary of criminality, http://www.eugenicsarchive.org/html/eugenics/static/themes/12.html.

13. Lewis Perry Curtis, *Apes and Angels: The Irishman in Victorian Caricature* (Devon: Newton Abbot: David & Charles, 1971), 13–14, cited in Steve Baker, *Picturing the Beast: Animals, Identity, and Representation* (Urbana: University of Illinois Press, 2001), 113.

14. Sir Everett Millais, *Two Problems of Reproduction: Lecture Delivered at St. Thomas' Hospital on February 25, 1895* (Manchester: "Our Dogs" Publishing, 1895), 23.

15. Millais, *Two Problems of Reproduction*, 23.

16. Derr, *Dog's History of America*, 61.

17. Eduardo Bonilla-Silva, "The Linguistics of Color-Blind Racism: How to Talk Nasty about Blacks without Sounding 'Racist,'" *Critical Sociology* 28, nos. 1–2 (2002): 26.

18. Bonilla-Silva, "Linguistics of Color-Blind Racism," 42.

19. *Dog and Kennel Magazine* is a bimonthly glossy magazine published by Pet Publishing Inc., Greensboro, N.C., http://www.petpublishing.com; *Doberman Digest* is a glossy magazine published ten times per year by Doll-McGinnis Publications, Lakeland, Fla., http://www.dmcg.com/pubs/contact_us.html; "Long and the Short of It," http://dachshundlove.blogspot.com/.

20. Mixed Up Mutts is a rescue organization in LaPorte, Indiana, run by Sarah Stevens and Cris Stevens, that places dogs in a prison training program, Prison Tails, http://www.mixedupmutts.org.

21. Jill Mosteller, "Animal-Companion Extremes and Underlying Consumer Themes," *Journal of Business Research* 61 (2008): 515.

22. Eric Konigsberg, "Beloved Pets Everlasting?" *New York Times*, December 31, 2008, http://www.nytimes.com/2009/01/01/garden/01clones.html.

23. Susan McHugh, "Bitches from Brazil: Cloning and Owning Dogs through the Missyplicity Project," in *Representing Animals*, ed. Nigel Rothfels (Bloomington: Indiana University Press, 2002), 187.

24. American Kennel Club, *The Complete Dog Book*, 20th ed. (New York: Ballantine Books, 2006), 52.

25. Readers are encouraged to see the Cottonelle Puppy TV ad "Cottonelle toilet paper Double Roll vs. the leading brand's Red Pack Big Roll," http://www.cottonelle.com/ads.aspx, from the Be Kind to Your Behind series of ads, Kimberly-Clark Worldwide, Inc., running in 2009 and 2010. Also see http://www.youtube.com/watch?v=4FFK5Oh10os for the video of DMX's 1999 song "What's My Name?" and http://disneydvd.disney.go.com/beverly-hills-chihuahua.html for "Beverly Hills Chihuahua" trailers.

26. Mark Doty, *Dog Years: A Memoir* (New York: HarperCollins, 2007); Elizabeth Marshall Thomas, *The Social Lives of Dogs: The Grace of Canine Company* (New York: Simon and Schuster, 2000); Jon Katz, *The New Work of Dogs: Tending to Life, Love and Family* (New York: Villard, 2003).

27. "Pit bull" is not a designation officially recognized by the AKC. However, there are a number of related breeds and mixes of those breeds—the American pit bull terrier, American Staffordshire

terrier, and the Staffordshire bull terrier—commonly referred to as pit bulls. Here I am using that vernacular term to mean those dogs that visibly resemble these AKC-recognized breeds.

28. Wright Thompson, "A History of Mistrust: Having Trouble Seeing Why So Many Atlantans See the Michael Vick Case as Racial Conspiracy?" ESPN, http://sports.espn.go.com/espn/eticket/story?page=vicksatlanta.

29. DoVeanna S. Fulton, "Comic Views and Metaphysical Dilemmas: Shattering Cultural Images through Self-Definition and Representation by Black Comediennes," *Journal of American Folklore* 117, no. 463 (2004): 81–96.

30. Dareh Gregorian, "Rapper DMX Mixed Up in Canine Suit," Foxnews.com, http://www.foxnews.com/story/0,2933,122436,00.html.

31. Malcolm Gladwell, "Troublemakers: What Pit Bulls Can Teach Us about Profiling," *New Yorker*, February 6, 2006.

32. Web site and rescue organization, "Contrary to Ordinary," The Black Pearls of the Dog World, http://blackpearldogs.com.

33. These associations can be regional; the black dog logo of Martha's Vineyard's Black Dog Tavern has cult status.

34. Amrah Salomon Johnson, "Is the Bandit Back Disguised as a Chihuahua?" *El Tecolote* online, March 7, 2005, http://news.eltecolote.org/news/view_article.html?article_id=f337b681beb06d75abffc2d06a03e7f4.

35. Champ Clark, "From Stray Dog to Movie Star: Rusco Leaves the Shelter and Finds a Home in the Hearts of Millions," *People*, November 3, 2008, 76.

36. Andrew Moseman, "Thanks to His Own Popularity, Nemo Can't Be Found," blog on Discovermagazine.com, June 27, 2008, http://blogs.discovermagazine.com/discoblog/2008/06/27/thanks-to-his-own-popularity-nemo-cant-be-found/.

Interspecies Families, Freelance Dogs, and Personhood

Saved Lives and Being One at an Assistance Dog Agency

AVIGDOR EDMINSTER

MY ETHNOGRAPHIC INFORMANTS AT A NORTH AMERICAN ASSISTANCE DOG AGENCY say they can read each other's minds, have saved each other's lives, hear for one another, are members of the same family, and are business partners. The clients, assistance dogs, and volunteers at the agency have uniquely intimate and interdependent interspecies relationships, which they cultivate and cherish despite the pervasive assumption of absolute differences between humans and all other species. My research examines the intricacies of these relationships, which my informants feel are so critical to their well-being. I concentrate on the ways my informants understand and create that which is shared and unshared, sharable and unsharable, between them. These ideas in turn reflect and inform characterizations of personhood, whether full, partial, or negated. In my informants' lives, personhood and notions of sharability often run counter to, although they sometimes overlap with, human-other animal distinctions.

Though species lines are often treated as self-evident, my informants attribute spectra of physical, affective, and cognitive abilities to humans and dogs in various ways. Such attribution authorizes or undermines various kinds of shared identities and experiences and reflects contemporary ideas and anxieties about work, family, and other fundamental matters. Critical both to human-dog distinctions and the perceived efficacy of my informants' relationships are these implicit and explicit notions of the sharable and unsharable, understood in terms of particular abilities, and more phenomenologically, in terms of worlds of experience.

Given the foundational character of species distinctions for identity- and meaning-making, my larger work disengages from many anthropological approaches which are predicated on a priori distinctions between humans (as people) and all other animals (as not people). Such distinctions are an integral element of the matrix of issues and relations that I consider ethnographically. I am somewhat constrained, however, by language and other conventions that reflect these distinctions, so for clarity my linguistic choices often reflect the language of my human informants. Nevertheless, I pay careful attention to the seeming contradictions and paradoxes that reveal the underlying instability of human-animal distinctions.

Manifesting an intensifying concern with human-animal relations in the social sciences,

Paul Nadasdy and Eduardo Kohn, among many others, assert that anthropology should address interspecies relations in new ways.[1] Given the crucial importance of interspecies relations, Kohn calls for an "anthropology of life" that attends to the embodied semiosis of all creatures.[2] Nadasdy has recently challenged anthropologists to examine the boundaries of sociality instantiated by most anthropological theory. He points out that most Euro-American notions of sociality reduce human-animal relations to the metaphorical rather than the actual. Like Tim Ingold,[3] Nadasdy observes that for arctic hunters, for example, animals are not *like* people but *are* people; thus Nadasdy challenges anthropology's historical denial of such a possibility.[4]

Where Nadasdy suggests that we must take human-animal relationships seriously by not reducing them to metaphor, I suggest that metaphor itself does not have to be reduction. Terence Turner explains that, from a positivist perspective, the literal is considered truer because it is understood to be "non-creative" and so more "natural."[5] Only within such a context and one in which metaphor and literality are understood as antithetical could metaphor be seen as mere reduction. Approaching notions about both the metaphorical and the literal as ways of knowing, expressing, and representing helps to illuminate even subtler and more complex articulations of people's relationships and experiences. Precisely how dogs may be understood as "metaphorical" or "actual" people, or "like" and "not like" human people in various contexts and relationships, as well as how some humans may or may not be like other humans (or dogs), is of critical importance. Our lives are profoundly shaped by the particular ways in which we liken one thing or person to another and to ourselves.

These processes of "likening" are critical to our shifting notions of personhood and our bonds to one another. My informants' relationships uniquely draw on and shape possibilities regarding personhood and sociality. Given the geometric increase in the number and kind of assistance and therapy dogs and their work in recent years, including the introduction of seizure and diabetic alert and assist dogs, assistance dogs for children with autism, animal therapy programs in elder residences, reading to dog programs in schools and libraries, and the announcement a few years ago of potential cardiac alert dogs, my informants' relationships reveal the persistence and emergence of various inter- and intraspecies socialities as well as the character of critical dimensions and criteria underlying and constituting these relationships.

The agency describes its mission as borne from philanthropic concern for humans with disabilities and dogs, some of whom are adopted from euthanizing animal shelters. The agency describes its work as "creating relationships," which involve physical and emotional interdependence between beings who are commonly understood to be of fundamentally different kinds. My informants enact and shape changeful ideas about what humans and dogs as such can share cognitively, emotionally, and physically, and about differences between them, sometimes as unbridgeable caesuras that differentiate them as species. My informants' work and lives together illuminate the variability of human-animal distinctions and relations even in relationships as "specific" or "specialized" as those between assistance dogs and clients.

During an open house at the agency, the executive director told the audience how he founded the organization. At the age of thirty-five he decided that "there was more to life than making money." He was inspired by Helen Keller, who, he explained, had declared, "I may be only one—but I am one." This quotation highlights a number of central notions that my informants' work and words often echo. Keller's words underscore a dynamic tension between what is perceived as deficiency and what is perceived as efficacy. In this declaration, "being one" is an invocation of a singular existence; it highlights a privileged wholeness, autonomy, and corresponding efficacy. Keller's words describe an ideal relationship between "being one" (person) and "doing" as one (person)—that is, between existence, ability, and productivity. This ideal is articulated after acknowledging the limitations inherent in being "only one" (person) on which this very autonomy must be predicated. Thus, given the tension between both the limitations and the efficacy of the individual as such, the director's use of Keller's assertion reminds us of the inextricable relation between notions of individual agency and a priori collectivity and sociality. The director's speech foregrounds ideals regarding what one person can (or should be able to) do, and it implicitly evokes ideas of dependence, independence, and interdependence that figure largely in my informants' perceptions and experiences of themselves and one another.

Ideas about differences between what some volunteers and staff "can do" and what clients "can't do" help to mobilize and shape the work and identities of my informants. This is equally true, of course, about the comparative nature of the perceived abilities, deficiencies, and identities of the dogs. The agency screens a film about its work at its open house events. This film closes with a quote that brings some of the foregoing ideas into stark relief. The quote is credited to John Bunyan, the seventeenth-century preacher and author of *Pilgrim's Progress*. The audience is advised that "you have not lived today until you have done something for someone who cannot repay you." Here as elsewhere, "living" and "doing" are inextricable and meaningful only in very actively comparative relations with others. To live (at all) here is to do what someone else cannot (hence the impossible repayment) and to do so for them. And this living, as doing for, must be done each day anew and thus renewed. This invocation of impossible repayment is reminiscent of many other practices of extravagant giving in which the giver or host "bests" receivers by highlighting their inability to return the gift. Many anthropologists and philosophers, from Bronislaw Malinowski and Marcel Mauss to Marilyn Strathern and Jacques Derrida, have explored various ideas regarding the renewed instantiation of sociality objectified in, and enacted by way of, "the gift."[6] For the present purposes, it is important to note that at the agency, the givers are reminded by this aphorism that they gain life by the affirmation of their own abilities in contradistinction to those they give to, by doing for.

The executive director's use of Keller's words is a tutelary signaling that highlights the shifting meanings of full and partial personhood in various contexts as agency clients, volunteers, and staff enact them. There is some disturbing irony here in that Keller's words draw attention to the abilities, efficacy, and perceived singular autonomy of the director, himself understood to be "able-bodied." Given the often implicitly comparative backdrop of much of the agency's philanthropic work, that wholeness is often understood and evoked in comparison to the deficiencies, or lack of wholeness, manifested by the clients who, like Keller herself, are often understood to be "dis-abled," not merely "differently abled" as we all are.

The agency very pointedly asserts that human clients and assistance animals are "partners" and "teams." This well-bonded team, while making a sort of whole, is also, however, no longer in Keller's words "only one" person; the abatement of aloneness carries with it a positive force. But in an equally important counter note, by being part of a "team" a client is thus (multivalently) unable, perhaps, to say as Keller did, or the director does, "I am one." By becoming part of a dog-client team, the partners in that team are neither "only one" nor "one" at all. "The team" here is, after all, a combination of *two* beings. While the work of the agency's volunteers and staff is consistently framed by reference to the questionable or imperiled full personhood of the clients, as well as the variable status of the dogs, it is equally clear that no one's full personhood is totally and permanently secured, including that of the volunteers and staff. Personhood must be affirmed and reaffirmed continuously.

The unique power of assistance dog–client relationships is explained and evoked by reference to "the bond" between clients and assistance dogs. As a frequently used term, "the bond" refers to powerful, emotionally charged, somewhat inaccessible and foundational elements of these relationships. The term is used variously to describe a relationship, a shared world, and an emergent quality. In its characterization of the client-dog relationship, "the bond" marks out a coherent dimension of relating even as it may be understood to be opaque or resistant to rational explanation. As such, "the bond" acts as a meaningful reference that allows social space for the traffic of the otherwise mysterious and ambiguous elements in these human-dog relationships. The conceptual space of "the bond" allows for the very possibility of the kinds of otherwise potentially threatening physical, perceptual, and emotional intimacy between dogs and humans on which the agency's work and the client-dog partnerships are founded and encourage such as shared perceptual and cognitive work and deeply felt partnerships.

"The bond" between assistance dog and client is invoked and described to fully authorize the efficacy of the agency itself and its teams, as well as to authorize those elements of client-dog relationships that are most ambiguous and potentially unsettling. While there is much variety in how various cultural domains, relationships, and beings are characterized, there are some noticeable disjunctions between the official descriptions and explanations of dog-client relationships by the agency staff and those of the clients. For the most part, agency staff, while underscoring the power of "the bond," tend to accentuate the nature of the client-dog team as a work relationship. The clients themselves tend to elaborate on the emotional or familial nature of the dog-human relationships, sometimes in ways that the staff seems more loath to acknowledge.

The agency's spokespeople and outreach materials reiterate that "freedom" and "independence" are what characterize quality of life and full personhood and what the assistance dogs offer their clients. As one client explains of her dog partner, "This dog means life, and a full life at that." Before being partnered with her assistance dog she had "lost all hope." For other clients, their dog partners allay fears. The executive director tells the story of one of the very first hearing-dog clients at the agency, a seventy-year-old man who contacted the director after his wife died. He told the director that he had not realized "how much [his wife] had heard for [him]." The agency was not then in a position to help him, but some time later the man called back and explained that "a bunch of motorcycle guys had moved in down the street." He said that he was sleeping with a "loaded .357 under [his] pillow," and pointedly asked the director, "How long would it be before they realize there is a seventy-year-old living alone?" The agency soon after found a dog for the man at a local shelter, arranged for training, and ultimately placed the dog with him. The director reassured the audience that the man then "unloaded the .357."

By way of conclusion, the director told the audience that when that man died, he left money to take care of his dog partner for the rest of her life and to train another dog for someone else. The loaded gun dramatizes the danger of the situation and sparks fear in the listener. Interestingly, in other contexts the gun alone might be called upon to counter the man's vulnerability and to reassure the listener. Yet here as elsewhere, we are shown that the perceptual help, as well as the dog's particular companionship, saves lives. The dog brings about the unloading of the gun, and this signals release from fear and danger, vicariously experienced by the listener as well. The fact that the story ends with the man's death and preparations made for the care of his dog partner leads the listener to conclude that the danger and fear remained at bay. This story illustrates the kind of lifelong and overarching freedom from fear that the agency promises to deliver through the teams it creates.

Another client explains that she works "downtown," and thanks to her dog partner she is "not scared walking to [her] car." Going through her daily life with her dog partner makes her feel as though she "got an extra set of ears and eyes." This allows her the sense of strengthened perception and so, in effect, an enlarged perceptual self. The sociologist Clinton Sanders writes that "the guide dog is literally experienced as an extension of the blind owner's self."[7] Although Sanders's description does seem to fit at least part of this woman's experience, assistance dog–human relationships, as my research reveals, are not experienced only in terms of "the extension" of some preexisting personhood. The dog is not necessarily experienced as an extension of the client, but the relationship definitely seems to expand her sense of self and life. In this example and others, such as the woman whose life became "fuller," these partnerships do get described as filling up a kind of "negative space" made by various perceived deficiencies, and so the client's very personhood is enhanced and strengthened. In these examples we are shown the powerful connections and mutual reinforcement between "peace of mind" and filled-out personhood.

While the notion of ownership as Sanders uses it will be touched on shortly, I here return to the problematic and inspirational juncture of metaphor and literality. Just as determining whether dogs are "real" or "metaphorical" people rests on the definition of people, the same holds true for determination of selfhood in Sanders's assessment, including whether selfhood can perhaps be "shared" (both as held in common and as a common attribute or experience). Regarding notions such as personhood and selfhood, sensitivity to the "likening" processes helps to illuminate what realities we feel and think we share with others. It may be that the same objectification of the dog as primarily owned here allows Sanders to perceive "it" as a "*literal extension* of the . . . owner," as this implies a certain emptiness of "its" own selfhood. I would suggest that there are many ways in which the human and dog partners here extend and change each other's lives and modes of experiencing. Additionally, these are shared and reciprocal, and not neatly literal or metaphorical.

Many of the stories that the agency's staff, volunteers, and clients tell contain elements of fear and vulnerability out of which the protagonists are rescued by their human-dog partnerships. Additionally, it is not uncommon for the very rescuing of the dog and the salvation of the human to be presented as interconnected. For example, another client of the agency tells the story of searching for a service dog. Self-described as "totally blind," she had worked with guide dogs for years. Then, as she explains it, after a bad car accident she needed to use a wheelchair and was told that she "couldn't use a service dog [with her chair] if she was totally blind." At that point she says she had "given up." Fortunately for her, the agency's director of training kept seeing a particular dog as she made her rounds to local animal shelters, and one day, when the trainer accidentally dropped her keys, the dog picked them up. The woman concluded her story by saying that with the retrieval of the keys, "we were both saved." She punctuated her story with the culminating revelation that her dog partner had been "slated to be euthanized the next day."

This is the paradigmatic rescue story of the agency in which two beings' lives were saved *in their pairing*. The dog was saved from being killed; what the human was saved from is not explicitly stated. But the fact that she expresses the sure parity between herself and her partner clearly underscores that while what imperiled her was not perhaps an immediate physical death, it was something that she understands as equivalent to death. As the client describes it, in this one exchange between the dog and the trainer both the client and the dog were spared the terror of their further undoing. The client, through explanation of her "worsening" conditions and rejection by other agencies, describes a decomposition of her life. It is clear that this is not only physical but inescapably social as well. She presents her position as similar to her dog partner about to be "put down." As she explains, the agency "was willing to give us a chance when no one else was." Just as she explains her own and her dog partner's simultaneous rescue from peril, she also presents the two of them as being in the same position, as socially abandoned, as needing to be "given a chance." Again, while she does not explicitly describe what the nature of this "chance" really is, it ultimately involved working together. Their salvation, together, comes with their potential for certain kinds of abilities and "self-sufficiency." In turn, the absence of this kind of autonomy and capability signals and can also bring on lack of life, whether diminished personhood or physical death.

Dogs in animal shelters, especially euthanizing shelters, are in unquestionably vulnerable positions. Their quality of life, including freedom of movement and access to food, as well as their continued existence, is entirely dependent on human decisions, whether capricious, thoughtful, rationally calculated, or emotionally motivated. When dogs are adopted they are often seen as "being given a chance" to be a satisfactory "pet" or "companion" to someone and are thus simultaneously "given a chance" to live. In fact, being given a chance to prove oneself in one way or another to humans is the only chance they have to live. The fact that this client consistently asserts the parity of her partner's position and her own makes it clear that what was, and is, at stake for both of them was equivalent. While she was not "slated to be euthanized," this woman clearly felt her life was imperiled.

Given this client's sense of impending doom and the fact that she believed her life was at stake when her future companion proved herself at the shelter so that they could both be "given a chance" by the agency, it is clear that notions of independence and the efficacy of one's work is foundational for the kind of full personhood and quality of life that is held as ideal in these contexts. The nature of this independence, like much else discussed here, is variable and context-dependent, but it is precisely its shifting nature that allows some people to secure a sense of independence for themselves and for others to lose it. In the foregoing description, the dog's personhood was highlighted by being paired in equivalency with the teller's own. At the same time, the woman who told the story underscored her own sense of marginality, vulnerability, and existential dependency by equating herself with someone in a shelter who was to be killed because she was unwanted and unvalued. By expressing this equivalence between herself and her partner, the woman makes clear that she and her partner have shared and continue to be able to share various kinds of experiences and social positions, though certainly not all.

The relationships between clients and assistance dogs are understood through reference to various kinds of important cultural domains, such as family or work. These are often drawn on in combination. In a single discussion a dog may be described as "like a son" and as "advertising himself" for assistance work or as undergoing a "career change." In all such examples the cognition, will, and capabilities, as well as the sociality of the dogs, are highlighted. At other times the dogs are discussed in ways that objectify them or highlight their lower status in a distinctively anthropocentric hierarchy. This is expressed when a trainer suggested that the volunteers bring their dogs into what is seen as a passive and subordinate physical position by "bring[ing] them down like a steer," to "let them know who's boss," or when a staff member

Figure 1. The assistance dog agency is home to a unique interspecies community. The agency's training classes are central to cultivating the skills needed in living and working together. (Source: Wikimedia Commons, http://commons.wikimedia.org/wiki/File:FEMA_-_15674_-_Photograph_by_Jocelyn_Augustino_taken_on_09-16-2005_in_Louisiana.jpg)

very pointedly reassured visitors on a tour of the agency that they "keep human and dog stuff separate" in the agency's kitchen.

So while in some of these contexts dogs are treated as persons, in others they are not. Such is the case when articulation of a client's newfound "independence" working with a service dog obscures the *inter*dependence between the client and his or her dog partner on which this "independence" is predicated. Given the pronounced chagrin with which one woman said that she "used to have to ask strangers for help," it is clear that this kind of dependency is viewed with fear and some shame. Working with her dog-partner, she makes clear, allows her "more independence," and yet presumably she is only independent to the extent that she is less reliant on *human people*—that is, in some very context-dependent sense, people like herself. Her relationship with, and reliance on, her assistance dog partner in this particular context is not acknowledged as the kind of "partnership" in which the dog's agency, personhood, or separateness would count in constituting an interdependency or a cooperative endeavor. Additionally, this new "independence" is considered a radical improvement in her quality of life, and so she becomes "more human" through her relationship with a dog. In this case, however, the dog's existence as a separate being must thus be negated, overlooked, or subsumed into the woman's own.

The status of dogs as at least partial "outsiders" to the full arena of personhood is, of course, foundational to the very possibility for the existence of "assistance animals" at all, at least as opposed to "assistants." However, the agency's staff, clients, and volunteers consistently

Figure 2. The agency often adopts dogs from shelters. They may then train to become hearing assistance dogs. (Source: Wikimedia Commons, http://commons.wikimedia.org/wiki/File:Dog_at_shelter.jpg)

assert that the dogs "want to do this work" and moreover that they *have to* want to do this work. Staff, clients, and volunteers reiterate that independent motivation, action, and desire on the part of the dogs are absolutely essential. As one client put it, "The dog has to want to do what you need it to do," and as the primary trainer says, "You can't make a dog do this." The kind of independence that is vaunted by the agency's staff, clients, and volunteers is precisely the kind that a dog in an animal shelter does not have. As the aforementioned client explains of her life before and after being paired with an assistance dog, "I used to have to ask strangers for help . . . [now] I have more independence." Dependence, especially on "strangers" (such as "biker guys" and anonymous denizens of "downtown"), looms in the background as a profound and ubiquitous source of peril.

The ability to work, on the other hand, is portrayed as the source of security and fulfillment. The dogs' relationships with their own work are framed as "careers." These are characterized as freely, individually, and independently entered-into life paths of skill-building and personal attainment that may include service provision for the agency's clients. In fact, when explained as part of the dogs' "careers," the provision of service is thus understood not only as something the dogs have chosen to do but also something that they continually choose to do. Of equal importance, this notion of a career is understood to confer benefits on the dogs, including presumably the satisfaction of a job well done and a career they thus enjoy. The agency's volunteer training manual contains side-by-side lists of "correct" and "incorrect" terminology for referring to the participants in their programs. "Client" and "team member," for instance, are "correct," while "recipient" and "owner" are "incorrect." Also included in this list is the declaration that "career change" is the "correct" phrase for describing a dog who has gone through the training program and who for whatever reason was deemed unsuitable as an assistance dog.

"Career change" thus stands as the proper description, while "flunked" is listed as "incorrect." Many of the agency's volunteers live with "career change" dogs. The precise nature of their "careers" after "career change" is not always clear but conceivably could involve being understood as a "companion animal," a "family member," or perhaps even a "pet." "Career" terminology thus frames the perceived unsuitability of the dog for assistance work as a choice made *by the dog* herself or himself rather than as a judgment made solely by the agency's staff.

The importance of what is understood as dogs' choices and desires is reflected in a former director of training's statement that when meeting and evaluating potential assistance dogs, she "look[s] for the dog who advertises that—that they want to help a human." The dogs' careers are characterized by independence, free choice, and personal reward. "Career" thus highlights professional identity and does not carry the same associations of laboring, constraint, and financial necessity as "work" or "jobs" do. Additionally, the use of "career" terminology elides potential association with other historical frameworks for understanding "working dogs." To an extent, this suggests a higher status for assistance dogs, whose careers are based on higher-status skills and are thus professionalized. This kind of "career" terminology also echoes the contemporary rhetoric that Zygmunt Bauman, Cosmo Howard, Richard Sennett, and others explain helps to craft all workers in neoliberal ordering into "independent" or "free agents" seemingly unloosed from previous social and institutional arrangements.[8] The status of a guard dog or herding dog relative to that of a hearing dog or seizure alert dog is perhaps comparable to what a bouncer or farm hand is to a paramedic or nurse social worker. It is important to keep in mind that the help assistance dogs provide is understood to be precisely what their human partners by definition *cannot* do. This is in marked contrast to many, though not all, other kinds of "working dogs." Though many "working dogs" may do work that humans do not want to do, such as guard property vigilantly or carry waterfowl in their mouths, many assistance dogs do what the humans they work with may very much wish they were able to do but cannot.

While human personal care attendants might seem a good human analogy for assistance dogs, this analogy would be misleading. It is precisely the dogs' perceptual skills rather than what gets understood as physical laboring, such as feeding and bathing a client, that renders this particular analogy inaccurate. Assistance dogs may be trained to carry objects, open doors, retrieve, and do many other physical tasks, but their work and the perception of their work far exceed these acts. The fact that their abilities and their own experiences are understood to remain elusive to human understanding may in fact help to boost their status as "professionals." After all, the unaided ability to hear outside of human range, somehow anticipate seizures, smell blood-sugar levels, or calm the inconsolable are not abilities most humans would lay claim to. These skills carry a relatively high (if also ambiguous) status, and thus in this context these skills are "professionalized" as they emerge as acknowledged possibilities in worlds of interspecies sociality. But because these abilities also bear an aura of the superhuman, they reemphasize some of the perceived differences between humans and dogs as very different kinds of beings.

In these very differences we can glimpse a second meaning for the phrase "you can't make a dog do this work." The abilities of assistance animals, while understood to be elicited and made consistent in training, are also understood to initially inhere in the dog, to be part of her or him, at least as potential. The dog must "want" to use that potential. Additionally, the assistance of the dogs is considered absolutely critical to the safety and well-being of the clients on a daily basis. Thus the dogs' own perceived level of independence is inextricable from their status as skill-bearing "professionals," just as the "independence" they provide for the human clients is ultimately what confers a fuller personhood on the clients as well. Lastly, the ability to have "careers" becomes a sharable element of life for humans and dogs, and of course for the clients and the dogs, their careers are often mutually dependent, highlighting their status as a "team." For many of the agency's clients, having a career and the social status this confers is practically impossible without an assistance dog's help, and certainly these dogs would not have these particular "careers" without the humans with whom they work. Career independence and success are essential to the ideal of full personhood that my informants at the agency so often describe.

Volunteers often speak about the nature of "clients" generically. A volunteer who has facilitated puppy-training classes and has fostered several dogs explained to a woman who was gesturing with her hands as she was calling the dog she was working with during a class that she should refrain from such movements because she "can imagine the client as not being able to use hand signals." While there is clearly a great deal of variety in the needs of the agency's clients, here we can note the practice of understanding, in fact "imagining" the clients generically as unable to do what the volunteers can do, in this case gesture with their hands. A volunteer puppy-raiser stated that though she finds it hard to let go of a puppy whom she has been training and fostering when the dog is placed with a client, she just reflects on the clients and how "they cannot do what we can do." Here clients are defined by some amorphous lack of ability. The criteria defining the "us" here are shared at least partially by the dogs (hence their work) as well as the volunteers, but *not* the clients. Whereas highlighting a client's increased "independence" while working with a dog reveals a temporary reinscription of boundaries separating humans from dogs, this volunteer's statement tacitly elides such a distinction. Here those abilities shared by the volunteers, assistance dogs, and generic others who are deemed whole give rise to a shared social identity that excludes at least the generic or hypothetical "disabled" clients of the agency.

Another puppy-raiser says of her experience fostering and then saying good-bye to a dog who is placed with a client that "it would be a waste of it if I kept it. . . . I am giving the dog to them." The "it" here is presumably a dog with whom she had formed some relationship, yet in this context the dog is objectified as a gift, in this case a bundle of skills involving a great deal of work on the part of the puppy-raiser. This work invested in the dog is passed on through training to the client. It is this work, or some kind of work, that the client is assumed to be unable to do without assistance. The dog's very existence, here understood as the dog's skills, would be "wasted" if the dog stayed with the foster-raiser because, as she makes clear, she does not have the needs that "they," the generic clients, do. This not only speaks to the perceived position of the clients as a group but again speaks to the dogs, whether objectified or endowed with personhood, as bearers of special skills. It is widely acknowledged that the dogs have skills and that these should not be "wasted." Like the client who along with her partner was "given a chance," the dogs and clients of the agency are consistently seen as being in the tenuous position of needing their potential worth acknowledged and their lives redeemed or salvaged by those in better positions.

Donna Haraway, in her very important work exploring how we can be ethically robust "companion species," asserts that "many of the serious dog people I have met . . . emphasize the importance of jobs that leave [dogs] less vulnerable to human consumerist whims."[9] She explains that many working dogs may or may not be loved, but their skills are valued; they are "respected for the work they do." Haraway maintains that "the status of pet puts a dog at special risk in societies like the one I live in—the risk of abandonment when human affection wanes, when people's convenience takes precedence, or when a dog fails to deliver on the fantasy of unconditional love." For Haraway, then, a working dog's "value—and life—does not depend on the humans' perception that the dogs love them"; the dog's life thus is not dependent on what she calls "an economy of affection."[10]

Accordingly, working dogs' lives then depend "more on skill" and "less on a problematic fantasy."[11] Here Haraway insightfully illuminates the position in which many dogs adopted by the agency find themselves. And her elucidation of "an economy of affection" is incisive. While many of the dogs who were "saved" from shelters by the agency are reflected in her description of the precarious position of being a "pet," it is also possible to hear in her words an echoing of the same sense of vulnerable dependence on strangers and being at the mercy of the convenience of other humans that clients of the agency have felt as well. Rather than their being cared for or remaining alive being dependent on their being loved, Haraway's working dogs, the agency's assistance dogs, and the agency's clients are all presented as having escaped a vulnerable dependency through skillful functioning, often framed in terms of maintaining careers. In the case of the clients, skillful functioning may be seen as manifesting a level of self-determination that is not necessarily measured by actually having a job or pursuing a career. However, the ability to do so remains embedded in frameworks for evaluating individual worth such as these.

Self-determination as measured by independence from the necessary help and caprices of other humans is considered an essential component of full livelihood for many clients. It is also clear that any "economy of affection" that dogs may find themselves dependent on is equally dangerous, if not more so, for humans who feel reliance on it. Such is hinted at by the man who believed that for his neighbors, the mere knowledge that he was living alone would impel them to harm him. In an "economy of affection," that dangerous "fantasy" that Haraway describes, a dog's safety is ensured only as long as the human believes that the dog loves him or her. A similar sense of the insecurity of any economy of affection binding humans in exchange pervades the words of many of the agency's clients. What may underlie the sense of peril that many clients feel in any dependence, especially on strangers, is the fact that there is no necessary expectation of affection or care from other community members at all. Real and abiding safety for dogs, according to Haraway, is to be found in productive work in what she characterizes as being a "functional dog."[12] I would argue that the same is conveyed by and about the clients of the agency and many others.

Haraway says of the approach of writer and sheepdog trialer Donald McCaig that "it might properly be called love if that word were not so corrupted by our culture's infantilization of dogs and the refusal to honor difference. . . . Dog naturecultures need his insistence on the functional dog preserved only by deliberate work-related practices, including breeding and economically viable jobs."[13] Here a dog's value and very life, as she has already made quite clear, is securable only through a "functionality" that is defined by being able to compete within a job market in a shared economy with humans. Haraway concludes that "we need . . . McCaig's knowledge of the job of a kind of dog, the whole dog, the specificity of dogs. Otherwise, love kills unconditionally both kinds and individuals."[14] By "kinds," Haraway is referring to breeds and/or types of breeds, such as herding dogs or game dogs. The "whole dog" also acknowledges "the herding dog" as a type transcending but characterizing the individual herding dog. The "whole dog" as she uses this term refers to what a given dog can do or who he or she is. A dog is, in her conception, not merely a "bundle of skills." Yet for Haraway, as for the puppy-raiser, a dog's capabilities can be "wasted." Accordingly, these skills and the jobs they are suitable for are the only abiding source of security for dogs as a species, as kinds, and as individuals. Breeds are thus at least partially critical to her vision of this security because they are the outcome of generations of breeding and skill-honing and so help ensure a competitive edge for dogs in a shared economy with humans without the need for recourse to "the fantasy" of affection.

While the "careers" of some assistance dogs seem to reflect a good deal of what Haraway

Figure 3. Many clients of the agency describe their relationships with assistance dogs as family. (Source: Ariel Cafarelli, used with permission)

explains and advocates, the experiences of the assistance dogs placed by the agency as well as the experiences and perceptions of the clients, staff, and volunteers help to underscore that separating "economies of affection" from other economies may not be possible, let alone ideal. Additionally, opposing "the reality" of the socioeconomic security of jobs or the possession of skills to "the fantasy" and ultimately murderous as well as infantilizing nature of something one might call "love," as she characterizes it, is at the very least quite problematic. The same kind of "bond" that clients, staff, and volunteers described as essential to a truly efficacious human-dog partnership is that which exceeds the framework of the job market or paradigm of the "career." While assistance dogs are clearly not solely dependent on "an economy of affection" in the same way as a "pet" might be, the various ways that the relationships between assistance dogs and clients are explained makes any clear distinction between "economies of affection" and skillful work an uncertain proposition.

If a client views a dog as "a member of the family" or "like a son," the familial and emotional dimensions of these relationships exceed the strict separation of work, love, and family that Haraway's framework seems to both take for granted and advocate. Haraway's argument regarding the potentially deadly nature of so-called affection and the place of dogs in a "fantasy world" of human self-absorption and its "pets" is important. However, it is by no means self-evident that the world of "functionality" and its criteria is any safer for anyone of any species. Assistance dogs are still ultimately dependent on humans whether they are evaluated by how well they do their work as assistance animals or by other criteria. The fact that they may well be more needed by those they work with is important, but neither does this fact guarantee their well-being.

The traps and dangers of conditional security are perhaps equally present for assistance animals as for dogs whose livelihood is secured by remaining pleasing as "pets." This is not to suggest that there are not important differences but rather to point out that there are significant similarities. While assistance animals as "working dogs" perhaps do not face the same threats of "outsourcing" and layoffs as other kinds of workers, like all kinds of other workers, canine and human, their skilled service provision does not permanently secure their well-being or their ethical treatment. Additionally, the notion of "functionality" is an amorphous yet potent characterization that renders the clients of the agency, and many others of us, lacking or partial.

The sociologist Leslie Irvine writes of her fieldwork at an animal shelter that people adopting dogs and cats often spoke of "the connection" that they felt between themselves and the dog or cat that they were bringing home.[15] She explains that she puts the phrase in quotation marks because it was the phrase consistently used by her informants, but also because she sees its common usage as "demonstrat[ing] the existence of an emotional vocabulary pertaining to interactions with animals."[16] Irvine explains that this emotional vocabulary is part of the larger "emotional culture," and she asserts that "the contemporary American emotional vocabulary includes rather traditional and straightforward words such as 'love' and 'anger,' but it also includes newer terms, such as 'freaked out,' 'stressed' and 'blown away.'"[17] Thus she says, "As new emotional states emerge within a culture, new vocabularies arise to describe them. 'Road rage' is an example. The notion of a 'connection' with animals is likewise a reflection of a particular time and emotional culture."[18] I would suggest that connections such as these have only expanded and multiplied since Irvine wrote her book. For more discussion on how connections with animals are situated in time and emotional culture, see Benjamin Arbel's chapter on the reconfiguration of animal meaning in the Renaissance in this volume.

Irvine does make clear that the perception that "the animal liked them" was essential for the adopter's feelings of "connection." "The affection signaled a 'connection'—or at least the potential for one."[19] Irvine's "connection" here is markedly similar to my informants' use of the phrase "the bond." And I concur that its consistent use speaks to emergent emotional states. In the case of "the bond" between assistance dogs and the agency's clients, these emergent emotional states are perceived to be fundamental for clients' well-being, as we have seen. But while there are obvious and important similarities between what Irvine describes as her informants' experiences of "the connection" and my informants' expressions of "the bond," there are important differences as well. Where Irvine highlights the perception that a dog or cat "likes" the human in question, my informants' discussion of "the bond" is understood as a fundamentally reciprocal dimension of the relationship, as well as one that is not limited to affection but involves navigating daily life together, physically, practically, perceptually. "The bond" is also not an immediately given experience. "The bond" may initially form more quickly or more slowly, but it is understood to grow and strengthen over time.

A visitor to the agency asked one of the clients during a question-and-answer session, "How long did it take the bond—the chemistry—to form?" The client answered without any need for clarification, "A year or so . . . until I know what she is thinking and she knows what I am thinking." This statement makes it even clearer that "the bond" is understood as an ongoing mode of or channel for relating in which there is an inextricable connection between mutual regard, emotion, and ability. Here again the human client and assistance dog share experiences, in fact skills, in tandem with one another that not only allow the client "a full life" but also, by highlighting the dog's cognition and mental world, acknowledge the dog's personhood. This statement defies a definition of personhood as the exclusive domain of humans. It also clearly points to abilities (of both a human and a dog) that exceed many standard working definitions of the capacities that humans, let alone dogs, possess. Here both dogs and humans are not only capable of thought but also of "knowing" one another's thoughts, regardless of, or exceeding, verbal communication.

The client's description of "the bond" here describes an experienced separateness as well as a shared reality between herself and her partner. Neither the dog's thoughts nor her "ears and eyes" belong to the client. They are not described here as an extension of her own. The client speaks of separate but sharable thoughts and a reciprocated "knowing" of the other. This description rests on, and attests to, the fact that her dog partner has thoughts and experiences, a mental as well as a physical existence equivalent to her own. The very full personhood of her dog partner that she attests to is concomitant with the regard she feels from her dog partner. She likens her partner to herself, and this testimonial to shared and sharable mutual regard defies many human-animal distinctions predicated on the human exclusivity of cognition.

Vicki Hearne, in her powerful and incisive writing on interspecies relationships, describes the process of building such relationships as becoming meaningfully coherent to one another.[20] Donna Haraway explains that one of Hearne's lessons is to make us aware that what some consider "philosophically suspect language" such as humans talking to nonhuman animals "is necessary to keep the humans alert to the fact that somebody is at home in the animals they work with."[21] For Hearne, language use while working with nonhuman animals is a way of maintaining an awareness of what Haraway might call "significant otherness" rather than an activity that assumes the dog or horse understands fully the language being spoken or understands it in the same way another human speaking the same language might.[22] Responding to Hearne, Haraway asserts that "just who is at home must permanently be in question. . . . The recognition that one cannot know the other or the self, but must ask in respect for all of time who and what is emerging in relationship, is the key."[23] The client who shared thoughts with her partner and many others at the agency clearly hang their livelihoods on such open-ended asking, just as their own fears of being preyed upon and others' dis-abling descriptions of them speak to the perils of not being equally regarded.

Haraway advises that all "ethical relating, within or between species, is knit from the silk-strong thread of ongoing alertness to otherness-in-relation. . . . *We are not one*, and being depends on getting on together. . . . The obligation is to ask who are present and who are emergent."[24] Many of the dogs and humans partnered by the agency manage their "getting on together" through relationships that defy any strict separation between the emotional and the functional, working and affection, family and career. These relationships and their "bonds" can also dramatically undermine the atomism that underwrites many notions of personhood itself.

The opposition of deficiency/dependency and wholeness/independence, expressed by the executive director and many others, is (somewhat ironically) further troubled by the very relationships that the agency helps to give rise to. The mutual regard and the skill at navigating life together that grow between clients and assistance dogs enlarge the perceived personhood of both partners and so simultaneously render "independence" a somewhat less straightforward concept. The contingencies that we all face as "we get on together" are no more escapable by committing ourselves to the cultivation of marketable skills than relying on economies of affection. The sheer variety of kinds of relationships called upon to characterize the "the bond" between human-dog partners, as family, coworkers, and so on, manifest Irvine's emerging "emotional vocabulary" that is inextricable from emerging relational and social possibilities. These defy any necessary distinction between, or separation of, work and family, affection and skill, or human and animal. But perhaps more important, these relationships point to possibilities outside of a social safety and mutual regard anchored in the fantasy of atomistic "functioning."

Powerful notions of ability and independence are mobilized to authorize the personhood of the agency's clients, volunteers, and many others. These same ideas are at work in descriptions of the assistance dogs and those in training. The livelihoods and lives of the clients and dogs, as well as volunteers and staff, highlight specific ideals of autonomy and career that imply a certain kind of atomistic personhood. Yet the interdependence and creativity that emerge relationally from client–assistance dog teams belie precisely that "oneness" of atomistic life that the director celebrates, even as those very qualities may also be used to illustrate successful if also paradoxical "independence." The personhood of the clients as well as that of the assistance dogs is made in reference to various kinds of abilities and notions of autonomy and independence even while various interdependencies are actively and passionately cultivated.

The experiences and abilities that clients and assistance dogs are understood to be able or unable to share authorize or undermine notions of shared personhood (and vice versa). Such elements may include shared thoughts as well as shared careers or, on the other hand, the exclusivity of language or the inability to hold a tool or walk. My work points to further questions

Figure 4. Mutual regard is central to the strength of "the bond." (Source: Avigdor Edminster, author photograph)

regarding the changeful and interpenetrating constitution of the categories of "human," "dog," and "person" in various contexts and how ideas about these shape people's experiences and our always collective and interdependent lives. Additionally, these putatively sharable or unsharable dimensions of life are bound up with the very likening process that resists absolute distinctions between metaphor and literality to the same extent that it expresses and enables emergent social possibilities.

Terence Turner, following Cristina Bicchieri, asserts that "the difference between . . . what is recognized as straightforward reference and what is perceived as metaphor . . . is not essential but pragmatic or contextually relative."[25] In addition to illuminating various disagreements regarding nonhumans as people, this insight also helps us to approach the ways that dogs (and others) may or may not "really" be part of families with humans. As the work of Janet Carsten, David Schneider, Marilyn Strathern, and others underscores, much anthropology on kinship has historically crafted as "fictive kinship" any relationships that are seen to originate in ways not in keeping with dominant Euro-American heteronormative marriage-based sexual procreation.[26] This in effect most often characterized as (meaningful) "fictions" those family relationships that did not mirror the anthropologists' own cultural schema for kinship.

Schneider, Carsten, and Strathern remind us that from that perspective, "real" family is reified as reflecting and continuing "blood" relations. The symbolic and very culture-bound quality of this "blood" is rarely questioned. However, as Carsten puts it, "Conceived in its broadest sense, relatedness (or kinship) is simply about the ways in which people create similarity or difference between themselves and others."[27] These scholars contribute a great deal to expanding and affirming our recognition of the many ways in which people do indeed reckon family and important relations outside of those frameworks alone. We can build on these insights to understand interspecies families as well as the ways we "liken" one to another, as resistant to metaphor-literal oppositions.

When Haraway writes that "we are not one," she is ostensibly reminding us that we must be alert to others and "otherness-in relation." But she also says that we must always ask "who are present and who are emergent."[28] Her use of the plural verb in "who are" reminds us of the impossibility of the kind of autonomous singularity that the executive director mobilizes when he takes from Keller the assertion that "[he] may be only one—but [he] is one." The human-dog partnerships discussed here illuminate some of the emergent possibilities for multispecies personhood and sociality. These may also help to foreground the necessarily contingent and relational nature of our lives together, whether we are perceived as functional individuals or not. The mutual reliance and contingency that "dependency" denotes is inescapable for social beings. Our lives and livelihoods always "hang" on one another, as the Latin root *"penderse"* expresses it, whether we perceive that as something to be feared, ignored, or cultivated in creative and beneficial ways.

NOTES

1. Paul Nadasdy, "The Gift in the Animal: The Ontology of Hunting and Human-Animal Sociality," *American Ethologist* 34, no. 1 (2007): 25–43; Eduardo Kohn, "How Dogs Dream: Amazonian Natures and the Politics of Transpecies Engagement," *American Ethnologist* 34, no. 1 (2007): 3–24.

2. Kohn, "How Dogs Dream," 6.

3. Tim Ingold, *The Perception of the Environment: Essays in Livelihood, Dwelling, and Skill* (New York: Routledge, 2000), 91.

4. Nadasdy, "Gift in the Animal," 31.

5. Terence Turner, "'We Are Parrots,' 'Twins Are Birds': Play of Tropes as Operational Structure," in *Beyond Metaphor: The Theory of Tropes in Anthropology*, ed. James W. Fernandez (Stanford, Calif.: University of Stanford Press, 1991), 124.

6. Bronislaw Malinowski, *Argonauts of the Western Pacific: An Account of Native Enterprise and Adventure in the Archipelagoes of Melanesian New Guinea* (1922; repr., New York: E. P. Dutton, 1961); Marcel Mauss, *The Gift: The Form and Reason for Exchange in Archaic Societies* (1950; repr., New York: W. W. Norton, 1990); Marilyn Strathern, *The Gender of the Gift: Problems with Women and Problems with Society in Melanesia* (Berkeley: University of California Press, 1988); Jacques Derrida, *Given Time: I. Counterfeit Money* (Chicago: University of Chicago Press, 1992).

7. Clinton Sanders, *Understanding Dogs: Living and Working with Canine Companions* (Philadelphia: Temple University Press, 1999), 57.

8. Zygmunt Bauman, *Liquid Modernity* (Cambridge: Cambridge University Press, 2000); Cosmo Howard, "Introducing Individualization" and "Three Models of Individualized Biography," in *Contested Individualization: Debates About Contemporary Personhood*, ed. Cosmo Howard (New York: Palgrave Macmillan, 2007), 1–44; Richard Sennett, *The Culture of the New Capitalism* (New Haven, Conn.: Yale University Press, 2006).

9. Donna Haraway, *Companion Species Manifesto: Dogs, People, and Significant Otherness* (Chicago: Prickly Paradigm Press, 2003), 38.

10. Haraway, *Companion Species Manifesto*, 38.

11. Haraway, *Companion Species Manifesto*, 39.

12. Haraway, *Companion Species Manifesto*, 39.

13. Haraway, *Companion Species Manifesto*, 39.

14. Haraway, *Companion Species Manifesto*, 39.

15. Leslie Irvine, *If You Tame Me: Understanding Our Connection with Animals* (Philadelphia: Temple University Press, 2004).

16. Irvine, *If You Tame Me*, 107.

17. Irvine, *If You Tame Me*, 107–108.

18. Irvine, *If You Tame Me*, 108.

19. Irvine, *If You Tame Me*, 109.

20. Vicki Hearne, *Adam's Task: Calling Animals by Name* (New York: Vintage Books, 1982).

21. Haraway, *Companion Species Manifesto*, 50.

22. Haraway, *Companion Species Manifesto*, 7.

23. Haraway, *Companion Species Manifesto*, 50.

24. Haraway, *Companion Species Manifesto*, 50, emphasis added.

25. Turner, "We Are Parrots," 129.

26. Janet Carsten, *After Kinship* (Cambridge: Cambridge University Press, 2003); David Schneider, *American Kinship: A Cultural Account* (Chicago: University of Chicago Press, 1980); Marilyn Strathern, *After Nature: English Kinship in the Late Twentieth Century* (Cambridge: Cambridge University Press, 1992).

27. Carsten, *After Kinship*, 82.

28. Haraway, *Companion Species Manifesto*, 50.

Animal Meaning in T. S. Eliot's
Old Possum's Book of Practical Cats

STACY RULE

WHILE THERE ARE MANY STUDIES OF ANIMALS IN CHILDREN'S NARRATIVE FICTION, there has been relatively little serious treatment of children's poetry—something individuals in both the children's literary and educational communities have expressed disappointment over. Agnes Perkins regrets that "compared with articles on fiction of all sorts, or even on nonfiction, and biographical articles on authors, and pieces on pedagogical methods, the number of critical works concerned with poetry for children is minuscule."[1] More recently, Anita Tarr has observed, "Even within the field of children's literature, poetry for children has suffered unaccountably."[2] Indeed, when readings do address poetry about animals, it is usually embedded in a larger discussion of fiction, and poetic meaning is often sublated as poems are read in much the same way as stories. One of the objectives of this chapter is to read T. S. Eliot's poems in a manner more appropriate to their poetic form.

I have found no scholarly analysis of *Old Possum's Book of Practical Cats*, a gap probably attributable to canonical issues and the comparative absence of scholarship on children's poetry.[3] Jeanne Campbell and John Reesman have pointed out, "Despite the book's moderate success in the past and its great success today, *Practical Cats* has occasioned almost no critical comment."[4] Instead, these poems are usually "dismiss[ed] . . . as nonsense verse or children's poetry."[5] Notably, despite pointing to the absence of criticism about these poems, Campbell and Reesman avoid such an analysis themselves, instead focusing on the way in which the book informs Eliot's other work. This decision evidently stems from their belief that "all of the poems of *Practical Cats* are creations that entertain, and it is very important that they not seek to *do* anything else. . . . Their charm is their expression of [Eliot's] personality and their gentle jokes."[6] Margaret Blount makes brief mention of the book, interpreting the cats as humans apparently because they wear clothes.[7] It is not difficult to see Eliot's collection fitting in with the dominant (animal represents human) reading. Certainly these poems humanize cats. As another review summed up, "The beauty of [Eliot's] conclusion is his clear statement that now that you have read about cats you have really learned a great deal about people."[8] Indeed, beyond brief comments about the anthropomorphic leanings of Eliot's children's book, these poems are rarely if ever interpreted as doing something else. This neglect belies a view that to read Eliot's animal representations as anything other than emblematic of the human condition is an illegitimate enterprise.

Given the lack of serious scholarship on *Practical Cats* and children's animal poetry more generally, the following review discusses commentary on animals in the larger genre of children's fiction. Academic readers have argued the status animals occupy in children's stories is often highly ambivalent, a condition generally taken as reflective of cultural incoherence. Critics also discuss the parallel divisions between adult and child, human and animal. Here, order, class, and power relations come to the fore, with species hierarchy among the most widely addressed topics. Therein humanity's privileged status is reinforced or jeopardized depending on how clear a boundary is maintained between animals and humans. Many have argued that objections to humanity's primacy are usually highly controlled and ultimately subdued. Blount has read these temporary reversals as evidence of "guilt at what Man has done to animals made deeper by the knowledge that animals can never 'win.'"[9] Further, in *Talking Animals in British Children's Fiction, 1786–1914*, Tess Cosslett notes even literature espousing kindness to animals can be hierarchal, such as those buttressed by a belief that people should be kind to creatures to whom they are naturally superior.[10] Although human hierarchy is usually upheld, there are promising exceptions. Linking eighteenth-century stories to philosophical discourse about how to educate children, Cosslett claims some texts have clearly been shaped by Rousseau's opinion that the nearer children are to animals, the less corrupt they will be—a concept that dislocates human supremacy because it means animals have something to teach us. Similarly, she notes works where an "animal's-eye view" shifts perspective, challenging anthropocentrism and asserting animal agency.[11]

Some scholars focus on the way children's literature engages with contemporaneous theories. Robert Hemmings, for example, takes a psychoanalytic approach, arguing that both authors and adult readers experience a kind of Freudian regression whereby they inhabit an imaginative space and time of an impossibly perfect childhood.[12] Such idyllic representations become narratives of empire, rewriting child experience and excepting real children. Tellingly, the forbidden elements of childhood are linked to animality.[13] Yet imperfections are not wholly disposed of as key upsets "expose anxieties and distresses that mobilize[d] the nostalgia" in the first place.[14] Similarly other critics have linked animals in children's fiction to nineteenth-century worries about evolutionary theory. In an article titled "Child's Place in Nature: Talking Animals in Victorian Children's Fiction," Cosslett discusses three stories as they relate to the Victorian recapitulation theory.[15] Here, problems in the theory amount to ambiguity in the story, such as the uncertain position child characters hold as the intermediary link between human adults and animals. Not only does this render the primitive child stranded, but it also makes it difficult for such texts to assert humans are superior to animals when animals, too, are engaged in an upward mobility, bound for humanity. Nevertheless, human order is restored as animal voices are dismissed as fantasy and finally silenced. Scientific discourse also figures in Rose Lovell-Smith's analysis of *Alice's Adventures in Wonderland*.[16] In this reading, concern over Charles Darwin's theory manifests when animal characters displace human authority as Alice transforms in physical size and public status, becoming just another "fellow creature."[17] She is repeatedly an object of investigation for animal characters who question her identity or misjudge it—implicit challenges to human power. Yet here again such power is ultimately secured.

Finally, considerable attention has been devoted to animals as symbols and tropes in children's stories. Instrumentalizing animals in order to explain something about people has a long tradition (see figure 1). In *Why the Wild Things Are*, Gail Melson observes, "Poking fun at human frailties via talking animals decked out in human attire—the animal burlesque—is one of the oldest literary conventions."[18]

To name but a few more, animal characters have been metaphors for humans; educational

A bon Chat bon Rat.

Figure 1. Illustration of a moral lesson, retaliation in kind, as "A bon chat, bon rat" means "to a good cat, a good rat." Granville (Jean-Ignace-Isidore Gérard), Cent Proverbes, H. Fournier, Paris, 1845. (Source: Wikimedia Commons, http://commons.wikimedia.org/wiki/File:Grandville_Cent_Proverbes_page65.png)

tools to teach children about human social rules and roles; ways to explore one's "animal" side; and vehicles to escape or express resistance to a changing world. With some exceptions, Elliott Gose's *Mere Creatures: A Study of Modern Fantasy Tales for Children* interprets these stories along these human-centered lines.[19] He claims "the human being is really the concern of the [*Winnie-the-Pooh*] tales," just as *The Wind in the Willows* is "really about human beings."[20] Blount gives a great many readings of symbolic animals who have "taken the place of humans and act out human dramas."[21] Yet she seems to regret that these stories exploit animal figures, and she balances these readings with other texts where humanizing animal characters is resisted, even impossible. The horses in *Gulliver's Travels*, she writes, "are reasoning beasts . . . as far from the human kind as one can get."[22] Her concern for actual animals is further clear in reminders like, "No one knows what animals feel, they can only guess."[23] Similarly, Cosslett emphasizes the persistent presence of animals. She cites naturalist authors like Beatrix Potter who valued representations that were truer to the lives of real animals. And, like Blount, Cosslett points out a strand of nineteenth-century authors of animal autobiography whose representations were driven by a desire to improve animal welfare. Further, she suggests, even in the most ambiguous works animal characters are not wholly absented. Countering Steve Baker's totalizing claim that children's stories manipulate animals to their exclusion, she argues these stories are "manifestly 'about' animals. . . . The animal characters in the stories may be metaphors for slaves, women or children, but they are also important metaphors for animals."[24] Indeed, animal figures are in these texts, however they are deployed.

Yet despite how extremely prominent animals are in these works, some critics simply do not deal with it, underestimating or ignoring how animals shape these stories. Anne H. Lunden, for example, surveyed seventy-five literary journals from 1880 to 1900 for their critical comment on children's fiction.[25] Her analysis, which sought key and recurring characteristics in literary criticism, yielded no mention of animals. This signals critical neglect since we know animals abounded in children's literature of this period. Given this, if fictional animals are made into pure symbols, perhaps it is the readers of children's literature (and not the writers) who are to blame.

I want to hold on to Cosslett's notion that animals in fiction are manifestly bound up with issues of animality and suggest that the way children hear these stories often differs from that of adults. Melson claims strict anthropocentrism is a culturally conditioned, adult tendency. Citing E. O. Wilson's "biophilia hypothesis," she argues children have an innate bond with animals (see figure 2): "The emotions and personalities of animals, *real and symbolic*, are immediate to children *in the same way* that the emotions and personalities of people are."[26] Thus, not only are animal depictions a safe means of exploring human emotions and interactions, but they are also a way for children to consider animality. Further, because of children's natural attraction to animals as such, Melson claims, we might use children to locate inconsistencies adults have learned to live with, look past, and not consider: "Children grapple with a complicated, often contradictory, mix of social codes governing animals and their treatment. There are creatures incorporated as family members, stamped out as pests, saved from extinction, and ground into Big Macs." Capitalizing on children's unique perspective, she writes, "may be the place to begin" as we attend to ecological concerns.[27]

As a children's book, *Old Possum's Book of Practical Cats* addresses an audience uniquely willing to consider animal subjects as complex, distinct personalities. The book's depictions allow readers to envision the lives animals have apart from humanity. Yet such possibilities are often undermined by readings that "project through [animals] psychological concerns . . . reader[s] either cannot or do not wish to experience directly in human terms."[28] While I do not dispute the value of such interpretations, I am troubled by the comparative lack of alternate readings that would redeem animal texts from a strictly human-centered perspective. Such routine readings obstruct discourse about animal conditions and instead redirect significance back to humans, depicting them as the sole proprietors of meaning.

Denied their own significance, animals are freed up for human appropriation. In Randy Malamud's evaluation of animal poetry, he argues, "The most pronounced trope that undercuts the value of most animal poetry is a sense of imperial mastery over animals: they exist for us to use as we please, in our life and in our poetry."[29] The easy disposal of animal identity[30] in such poetry is matched by the grim status of real animal lives. For, as Malamud asks, "What is any animal after we are done with it—after it has served our cultural purposes,

Figure 2. A child hugs her cat. Léon Bazile Perrault, *Son Favori* (Her Favorite Pet). (Source: Wikimedia Commons, http://commons.wikimedia.org/wiki/File:Perrault_Leon_Jean_Basile_Son_favori.jpg)

had the culture sucked out of it? . . . [It] disappears: farewell, farewell."[31] He finds poems that merely describe animals less egregious, but still lacking since they do not call on readers to imagine animal perspectives. Similarly, poems written in the "excursion format," where speakers encounter animals on trips into nature, are problematic as they position animals and nature as separate from the human world and therefore relatively insignificant.[32] The poetry Malamud praises demonstrates the ecocritical belief that everything in the world is inherently connected. Accordingly, animal poetry ought to counter what he calls our culture's "speciesist discourtesy," marked by "the myriad of ways we mistreat animals, physically as well as metaphysically (aesthetically, intellectually, ethically)."[33] What I find especially compelling about Malamud's book is his perspective of animal poetry as an imagining, a way of engaging with animals at a safer distance: "Animals generally suffer whenever they come in contact with people, and I think this is why [Marianne] Moore, who wants to depict animals with as much integrity and dignity as possible, chooses a stance . . . at a discreet remove."[34] Granted, humans can (and do) interact with other creatures without harming them, but too often we do not, even when our efforts are genuine. Indeed, some of the poetry Malamud champions falls short of his standards, but such transgressions, he writes, are forgivable given what the poet is trying to accomplish and mainly achieves. Good animal poetry is self-conscious about the limits of possible knowledge. It recognizes that animals exist for their own sakes and in their own contexts and ways.

Practical Cats is consistent with Malamud's aesthetic. The first poem appearing in the collection, "The Naming of Cats," complicates the assumption that animal subjects are trivial. This initial address to readers attempts to regulate attitudes toward the project. The poem begins:

> The Naming of Cats is a difficult matter,
> It isn't just one of your holiday games. (lines 1–2)

It is clear that on some level "The Naming of Cats" refers to the collection as a whole (which is largely organized around naming), and suggests that while playfully presented, the poems are nevertheless complex. Here Eliot preempts two false assumptions: children's poetry is unsophisticated, and animals are simple beings. Eliot's sensitivity to readers' views about animal subjects is evidently justified. According to Campbell and Reesman, he did garner criticism from an old neighbor of his, Bertha Rives Skinner, who wrote to him asking why "in this big round world" did he bother to write about cats?[35] Glen A. Love has written, "It is one of the great mistaken ideas of anthropocentric thinking, and thus one of the cosmic ironies, that society is complex while nature is simple."[36] It is my contention that scholarship on animal poetry should counter this bias by presenting nuanced readings that unpack not just diction but also meter, rhyme, style, and other elements of poetic composition that contribute to meaning and distinguish it from prose. These "poetic tools" are things Malamud expressly values in *Poetic Animals and Animal Souls*, but regrettably does not pay much attention to in his critiques.[37]

The speaker's assumption of a wary reader who needs convincing is characteristic of Eliot's poetry. Louise Glück claims Eliot's speakers want "communion . . . attention secured."[38] Yet this harmony is withheld as "Eliot's speakers either can't speak or can't be heard; their persistence makes the poems urgent."[39] Indeed, from the outset "The Naming of Cats" seeks to counter the idea that naming cats is insignificant. This constructed opposition foregrounds the forceful declaration in lines 4 and 5 that "a cat must have three different names," while the urgency Glück writes of appears in the repetitive pleas, "But I tell you" and "I tell you" in lines 14 and 27, respectively. Far from a playful endeavor, the naming of cats becomes of utmost importance in this poem. As we will see, naming here has everything to do with claims of cat identity.

Naming always strikes a relationship. In "Naming as Social Practice: The Case of Little Creeper from Diamond Street," Betsy Rymes sketches what she hopes will serve as a foundation for a "social theory of naming," one that would help explain how names function and what meanings they offer in different contexts.[40] Building upon the work of several theorists, Rymes writes that names, at their most basic level, behave in two ways: "The name an individual is given has one synchronic meaning in the baptismal [name-giving] ceremony; but as the individual uses that name, it acquires new and varied meanings diachronically."[41] The first (synchronic) aspect of a name is concerned with the naming ceremony, where an individual receives a name and a relationship is formed. Thenceforth, when that name is spoken, it references both the named individual and the one who named him or her.[42] Once this relationship is created, the second facet of naming involves the association of the name with certain qualities related to the individual. Citing John R. Searle's phrase, Rymes writes that "names function 'not as descriptions, but as pegs on which to hang descriptions.' Thus the meaning of one's proper name evolves through a life history."[43] This is the process through which a name moves beyond pure index and becomes a meaningful label intricately bound up with selfhood. This helps explain why it is more difficult to sanction violence against an animal who has a name. This reality is clearly illustrated in Melson's account of 4-H programs in which children raise animals for slaughter:[44] "The goal is to learn to *care for* and *care about* the animals, even to recognize an animal's personality and delight in its quirky behavior, but then to be able to have this same animal killed and turned into its monetary value without regret."[45] Melson goes on to note that the most common way 4-H children negotiate this problem is to resist naming the animals. Such resistance is a distancing technique that helps block a relationship and denies each animal's identity.

"The Naming of Cats" is mainly concerned with causality, the "baptismal event" Rymes cites.[46] Indeed, each kind of name the speaker describes functions differently, drawing a different relationship. It is the first kind of names that are expected, general, and easy to say. We have heard these names before, just as line 13's "all of them sensible everyday names" is something we have already heard in line 9. This pattern gives a sense of banality and helps readers to stay ahead of the lines. Likewise, in line 11, "Some for the gentlemen," coupled with the rhyming scheme, more than prepares us for the predictable "some for the dames." Minimal attention is required. Significantly, because readers have been promised difficulty from the outset, what is easy to articulate becomes less interesting, less relevant, and weak. Similarly, line 12's names, "Plato, Admetus, Electra, Demeter," are already familiar and easily said. They come with histories, which might cause the cats to be associated with and weighed down by the qualities of original name bearers. This threat is averted by the second kind of name, which is concerned with particularity, something equated with dignity and pride. This name must "never belong to more than one cat" (line 21). Here the proverbial "peg" is given free of a history, new.[47] Eliot limits our ability to say such names like "Munkustrap, Quaxo," and "Coricopat" (line 19). We are meant to stumble over "Bombalurina, or else Jellylorum" (line 20). Unfamiliar, unique, and more difficult to pronounce, these names are deployed to slow us down and procure attention. The very difficulty and uniqueness of this second class of names seem to protect them from being shared by other individuals, thus guarding the particularity of the cats who bear them. The audience is likewise forced to carefully read the assertion that "a cat needs a name that's particular, / A name that's peculiar" (lines 14–15), for these lines are highly similar, yet not the same. Calculated to trip up a fast reading, one nearly repeats "particular" in line 15 and has to look again. The point here is slight difference. Despite apparent resemblance to lines 6–13, these middle lines demand more. We are not entirely permitted to fall back into the poem's original rhythm. We read similarly, but not quite the same way. This difference embedded in similarity is an appeal for cats. Indeed, lines 16 and 17 reference physical traits ("tail . . . whiskers"), a return to generality, only to soon suggest the difficult names referenced above. While alike in form, both these poetic lines and cats themselves are quite different in content.

In the end, it is precisely interiority we are left with. The third name the speaker describes is one we can neither "guess" nor "discover" (lines 23–24). It is linked to private thought and is fully guarded: "The cat himself knows, and will never confess" (line 25). Following this, the speed of the poem radically increases. Line 26 fixes the base rhythm, and the next two lines keep to it:

> When you notice a cat in profound meditation,
> The reason, I tell you, is always the same:
> His mind is engaged in a rapt contemplation . . . (lines 26–28)

These lines build and build, finally culminating in a communicative rupture: "Of the thought, of the thought, of the thought of his name:" (line 29). Here the repetition insists on cat's thinking, at once stalling readers and impelling the poem's argument. Likewise, Eliot's manipulation of colons elaborates but never wholly delivers a cat's "reason" or "name" (lines 27, 29).

Further, reiterating "thought . . . thought . . . thought" and "able . . . able . . . able" constitutes a second building, one that juxtaposes feline capability to think with human inability to know what it is they are thinking (lines 29–31). Delaying readers while advancing cats dislocates notions of human authority over animals. Finally, line 31 gives us "Effanineffable," a word that is not a word, but which nevertheless we are confronted with, unable to understand. This pattern of letters, certain parts of which resonate with what came before, is both engaging and disorienting. There is an unsatisfactory semblance and an impenetrable depth. At the end of this poem, we can access only a little of the cat condition. The rest is flatly denied us.

While naming provides the basis for relationships with and discourse about animals, on some level it is unavoidably an imposition. As Donna Haraway notes, language entails a power relation: "Indeed, it is said that language is the tool of *human self-construction*, that which cuts us off from the garden of mute and dumb animals and leads us to name things, to force meanings, to create oppositions, and so craft human culture."[48] As we name, we construct ourselves in relation to what we name, determine the perimeters of what we deem important, and formulate meaning. Haraway cites the classification of animals as an example: "Linnaeus's taxonomy was a logic, a tool, a scheme for ordering the relations of things through their names."[49] Here the name-giver creates the boundaries of species and decides what will constitute an animal's family. To recall Rymes's reference to the baptismal event, as a name "indexes a relationship as much as an individual," an animal's given name might be a constant reference to human power over animals.[50] Indeed, the linguistic construction of identity through naming is violent because it fuels the pernicious fantasy that we create what we name. However, one must remember that when an animal is named, it is the *relationship* that is created, not the animal. That is, a creature does not come into existence because one names it; it exists already. "The Naming of Cats" suggests this with the third kind of name, "the name that no human research can discover" (line 24). This attests to identity before the name. Eliot's speaker reminds readers what they cannot say or know and, therefore, what they can neither create nor control.

The assertion that cats have a private life, inaccessible to humans, is central to the book. In "The Old Gumbie Cat," for example, Jennyanydots is far more than the humans in the poem believe her to be. The refrain states:

> She sits and sits and sits and sits—and that's what makes a
> Gumbie Cat! (lines 5–6)

This assumes every Gumbie Cat is the same, and all of them spend their time sitting. But readers soon learn this description is *not* what makes *this* Gumbie Cat. While the family sleeps, she teaches and feeds mice, and she manages the resident cockroaches. She cannot be defined by what the family sees, for despite their estimation, she is responsible for keeping "a well-ordered house" (line 45). Following the logic of "The Naming of Cats," this poem situates its most forceful claims of feline particularity in the lines that break form. The odd-numbered stanzas give superficial information and are slow and drawn out:

> I have a Gumbie Cat in mind, her name is Jennyanydots:
> Her coat is of the tabby kind, with tiger stripes and leopard spots.
> All day she sits upon the stair or on the steps or on the mat: (lines 1–3)

By contrast, the second, fourth, and sixth stanzas violate the original pattern, featuring centered text, condensed syntax, and more lines with fewer beats. They speak to the cat's uniqueness—what she "is deeply concerned with," "decides," and is "believing." Form reflects her true complexity as these lines are faster, more energetic:

> But when the day's hustle and bustle is done,
> Then the Gumbie Cat's work is but hardly begun.
> And when all the family's in bed and asleep,
> She tucks up her skirts to the basement to creep. (lines 7–10)

Such illustrations of the private lives of cats help counter an aspect of naming that threatens to reduce an animal. Boria Sax has argued that "simply by saying the name 'squirrel' . . . we dissipate part of the mystery."[51] The speaker's emphasis on the failure of humans who so misjudge the cat in this poem restores that mystery. As Malamud notes in his critique of animal poetry by Marianne Moore, Eliot's contemporary, "The issue of names and misnaming is a way of showing people's clumsy inability to know animals, literally and precisely, amid our cultural proclivities to manipulate things and despite our fantasies of omniscience."[52] Accordingly, the human position in this collection is expressly unable to alter the behavior of these animal subjects (see figure 3). The refrain in "The Rum Tum Tugger" insists that this cat's "disobliging" ways cannot be changed:

> For he will do
> As he will do
> And there's no doing anything about it! (lines 9–11).

Figure 3. Cat surprised by the sound of the camera's shutter. By Olivier Aumage, 2005. (Source: Wikimedia Commons, http://commons.wikimedia.org/wiki/File:Pocket_tiger.jpg)

Likewise, in "Mungojerrie and Rumpleteaser," the confusion is permanent: "And there's nothing at all to be done about that" (line 31). This inability to control proceeds from each cat's private identity.

If we view the "The Naming of Cats" as a project about constructing relationships, then "The Ad-dressing of Cats" is one about the mining of identity. Contrary to the first poem, this one features a cat as an active participant who already has a diachronically functioning name, without the aid of the human speaker. This cat has an identity, an undisclosed name, and power. The speaker notes:

> With Cats, some say, one rule is true:
> *Don't speak till you are spoken to.* (lines 38–39)

Such a position would have humans waiting for permission to engage on the cat's terms. The speaker rejects this protocol:

> Myself, I do not hold with that—
> I say, you should ad-dress a Cat. (lines 40–41)

This advice signals a desire for an equitable relationship, not merely a reversal of the usual power structure. Indeed, the speaker criticizes relations marked by control and servility in his assessment of dogs in general. They are "very easily taken in" and far too subservient to humans (line 30). Rather than speciesism, dogs here serve as Eliot's trope for the animal pet humans want and consistently seek to shape—obedient, predictable, and unproblematic. Such a relationship does not challenge but instead enforces human superiority. Unless one is a "fantastic freak," in the speaker's view, wholesale tameness and blind submission make most dogs "simple soul[s]" (lines 25, 23). Accordingly, while the first poem celebrated the dignity and pride of cats, here dogs are expressly "far from . . . pride" and "undignified" (lines 28–29). Likewise, addressing cats is a complicated endeavor, but dogs will "answer any hail or shout" (line 35). Therefore the poem stresses, "A CAT IS NOT A DOG," and the speaker espouses certain rules for interacting with cats (line 19). He wants to address the cat next door in a way that would engender mutual consideration, directed by some combination of human and feline conventions. Of course, line 48's parenthetical "He comes to see me in my flat" suggests such an alliance is vulnerable to misinterpretations. That is, the cat may already have initiated a dialogue in its own nonhuman way, one the speaker does not recognize as such. Access to identity is a process: "But we've not gotten so far as names" (line 51). The speaker's true "aim" is to "finally call him by his name" (lines 66–67). This must be earned through careful and curious interaction, the basis for identity disclosure. In the meantime, the speaker suggests humans use temporary titles to talk about cats—"O CAT!" for strangers and "OOPSA CAT!" for acquaintances (lines 45, 49). The second title, a merging of "Oops, a cat," builds on the first title and memorializes the original encounter when the speaker stumbled upon the cat, the beginning of a relation, but not yet a revelatory exchange.

As in "The Naming of Cats," the layered names in this poem are against possession, but still bespeak connection. This distinguishes Eliot from poets who, as Malamud notes, take a stand against naming: "The singular poets are those who, in [David] Weiss's construction, 'refuse to name' . . . manage to evoke and poeticize animals without colonizing them and without constraining them in the methods and limits of our own knowledge system."[53] But this stance falls too much in line with separatist notions that divide nature and culture, for if we cannot speak of animals, how can they matter *to us* as they must if we are to embrace a more biocentric perspective? Eliot's manifold names counter isolation while at the same time resisting the fallacy of total comprehension of animal others.

Finally, in "Cat Morgan Introduces Himself," a poem not included in the original publication but added in 1952, the subject offers his own name. This plays out what the speaker of "Ad-dressing" hopes for—a personal introduction, a self-revealing. This point of contact fits Malamud's prescription for better animal poetry. Critical of works where animals begin and end in the poem, never to exist outside the text, Malamud proposes that "the conclusion of a poem should not signify the closure of the relationship between person and animal, but rather, ideally, should initiate and inspire the *beginning* of an imaginative consideration and reformulation of who these animals are and how we share the world."[54] This is precisely what Eliot's collection accomplishes. Its encounters are not confined to the poem, but are suggestive of worldly meetings (see figure 4). The poems offer advice about how readers might approach autonomous and internally complex animals and imagines how they might address us. The assertion of a separate identity, one that must be solicited, not given, supports an alternative reading of the claim made in "The Ad-dressing of Cats": "That Cats are much like me and you" (line 6)—for anthropomorphism can be, as Clinton R. Sanders and Arnold Arluke argue, a "useful heuristic device" with "the practical utility of allowing the construction of

Figure 4. A close-up of a cat's face. (Source: Wikimedia Commons, http://commons.wikimedia.org/wiki/File:Cat_face _closeup.jpg)

effective and mutually rewarding patterns of social interaction." Under this model, shared qualities counter the "psychological distance necessary to exploit animals ruthlessly."[55] Instead of prompting us to see cats as symbols for humans, Eliot calls for an equalization of the two species based on singular personalities.

Children's poetry is a rich medium for imagining, however fantastically, animal personalities and experiences. These works often feature animals acting with purposes born of individual motivations and desires. Ultimately, this representation reforms assumptions of humanity's limitless control. Further, since children (who often see animals as equals) are the most receptive to such ideas, animal poetry is a fertile arena for equipping future generations with values they will urgently need given the ecological problems they stand to inherit. Finally, turning to this genre is also a way to reform adult views by attuning to a perspective that once came naturally. Through children's poetry, we might inhabit the space that came before we "absorb[ed] a worldview of humans as radically distinct from and superior to other species."[56] As poems endow the culturally voiceless with voices, the way we read determines how (and whether) we hear them.

NOTES

1. Agnes Perkins, "Critical Summary of Recent Journal Articles on Poetry for Children," *Children's Literature Association Quarterly* 5, no. 2 (Summer 1980): 35.
2. Anita Tarr, "'Still So Much Work to Be Done': Taking Up the Challenge of Children's Poetry," *Children's Literature* 28 (2000): 195.
3. T. S. Eliot, *Old Possum's Book of Practical Cats* (London: Faber and Faber, 1981).
4. Jeanne Campbell and John Reesman, "Creatures of 'Charm': A New T. S. Eliot Poem," *Kenyan Review* 6, no. 3 (Summer 1984): 25.
5. Campbell and Reesman, "Creatures of 'Charm,'" 26.
6. Campbell and Reesman, "Creatures of 'Charm,'" 31.
7. Margaret Blount, *Animal Land* (London: Hutchinson, 1974).
8. Owen Kerr, "Old Possum's Book of Practical Cats by T. S. Eliot," *English Journal* 65, no. 1 (1976): 66.
9. Blount, *Animal Land*, 24.
10. Tess Cosslett, *Talking Animals in British Children's Fiction, 1786–1914* (Aldershot: Ashgate, 2006).
11. Cosslett, *Talking Animals*, 36.
12. Robert Hemmings, "A Taste of Nostalgia: Children's Books from the Golden Age—Carroll, Grahame, and Milne," *Children's Literature* 35 (2007): 54–79.
13. The state of being a nonhuman animal.
14. Hemmings, "Taste of Nostalgia," 5.
15. Tess Cosslett, "Child's Place in Nature: Talking Animals in Victorian Children's Fiction," *Nineteenth Century Contexts* 23 (2002): 475–495.
16. Rose Lovell-Smith, "Eggs and Serpents: Natural History Reference in Lewis Carroll's Scene of Alice and the Pigeon," *Children's Literature* 35 (2007): 27–53.
17. Lovell-Smith, "Eggs and Serpents," 2.
18. Gail Melson, *Why the Wild Things Are* (Cambridge, Mass.: Harvard University Press, 2001), 145.

19. Elliott Gose, *Mere Creatures: A Study of Modern Fantasy Tales for Children* (Toronto: University of Toronto Press, 1988). Examples of exceptions to Gose's otherwise anthropocentric readings include his assertions that *Charlotte's Web* is in part concerned with teaching interspecies affection and that *The Mouse and His Child* suggests a universal spirituality in all creatures.

20. Gose, *Mere Creatures*, 41, 43.

21. Blount, *Animal Land*, 25.

22. Blount, *Animal Land*, 63.

23. Blount, *Animal Land*, 43.

24. Cosslett, *Talking Animals*, 182.

25. Anne H. Lundin, "Victorian Horizons: The Reception of Children's Books in England and America, 1880–1900," *Library Quarterly* 64, no. 1 (January 1994): 30–59.

26. Melson, *Why the Wild Things Are*, 19, emphasis added.

27. Melson, *Why the Wild Things Are*, 20–21.

28. Gose, *Mere Creatures*, 5.

29. Randy Malamud, *Poetic Animals and Animal Souls* (New York: Palgrave MacMillan, 2003), 27.

30. The unique condition of a particular animal.

31. Malamud, *Poetic Animals*, 29.

32. Malamud, *Poetic Animals*, 33.

33. Malamud, *Poetic Animals*, 64.

34. Malamud, *Poetic Animals*, 96.

35. Campbell and Reesman, "Creatures of 'Charm,'" 29.

36. Glen A. Love, "Revaluing Nature: Toward an Ecological Criticism," in *The Ecocriticism Reader: Landmarks in Literary Ecology*, ed. Cheryll Glotfelty and Harold Fromm (Athens: University of Georgia Press, 1996), 230.

37. Malamud, *Poetic Animals*, 159.

38. Louise Glück, *Proofs & Theories: Essays on Poetry* (Hopewell, N.J.: Ecco Press, 1994), 21–22.

39. Glück, *Proofs & Theories*, 22.

40. Betsy Rymes, "Naming as Social Practice: The Case of Little Creeper from Diamond Street," *Language in Society* 25, no. 2 (June 1996): 237–260.

41. Rymes, "Naming as Social Practice," 239.

42. Rymes, "Naming as Social Practice," 245.

43. Rymes, "Naming as Social Practice," 239.

44. 4-H is a national program for youths funded by the U.S. Department of Agriculture. Its participants develop skills in science, engineering, technology, health, and citizenship. "4-H" stands for Head, Heart, Hands, and Health.

45. Melson, *Why the Wild Things Are*, 69.

46. Rymes, "Naming as Social Practice," 239.

47. Rymes, "Naming as Social Practice," 239.

48. Donna J. Haraway, *Simians, Cyborgs, and Women: The Reinvention of Nature* (New York: Routledge, Chapman, and Hall, 1991), 81, emphasis added.

49. Haraway, *Simians, Cyborgs, and Women*, 81.

50. Rymes, "Naming as Social Practice," 247.

51. Boria Sax, "Animals as Tradition," in *The Animals Reader: The Essential Classic and Contemporary Writings*, ed. Linda Kalof and Amy Fitzgerald (Oxford: Berg, 2007), 271.

52. Malamud, *Poetic Animals*, 129.

53. Malamud, *Poetic Animals*, 58.

54. Malamud, *Poetic Animals*, 33–34.
55. Clinton R. Sanders and Arnold Arluke, "Speaking for Dogs," in *The Animals Reader: The Essential Classic and Contemporary Writings*, ed. Linda Kalof and Amy Fitzgerald (Oxford: Berg, 2007), 67–68.
56. Melson, *Why the Wild Things Are*, 20.

Animals at the End of the World
Notes toward a Transspecies Eschatology

CASEY R. RIFFEL

FROM THE INITIAL SCENE OF PHILIP K. DICK'S DYSTOPIAN SCIENCE FICTION NOVEL *Do Androids Dream of Electric Sheep?* (1968), issues of empathy, commodity fetishism, and ontological authenticity frame how questions of human and animal being play out in a postapocalyptic world. Deckard, one of few humans remaining on Earth after an apocalypse—ambiguous in its cause if not in its effects—emerges onto the roof of his apartment building to attend to the electric sheep he keeps as a pet. Attending to its small patch of grass, the sheep literally embodies humanity's anxiety about its uncertain relationship to the past; the keeping of ersatz animals immediately links the human-animal relationship to biological and historical nostalgia. So, from the opening chapters of Dick's novel, we must consider the question of how the ability to manufacture superficially if not subcutaneously convincing animals and androids alters the stakes of the human use of animals. The popularity of Dick's novel (hereafter, *Do Androids Dream*), owing as much to the canonical status of its film adaptation, *Blade Runner* (dir. Ridley Scott, 1982), as to its perspicacity, means that the story of Deckard as android bounty hunter and his subsequent development of empathy for "electric animals" is a particularly useful entry point for this discussion. It seems that the question, especially in the context of this volume, becomes not the making of animal meaning, but the meaning of making animals. That, for Dick, this inversion must be answered in terms of empathy is nothing new for those familiar with the reception and criticism of the novel. However, despite the salience of the ersatz nature of the electric animals, the novel ultimately and productively evacuates ontology from the task of recognition. Despite the pervasive anxiety around the covering and uncovering of any given animal's identity as "real," what comes, in an apocalyptically "uncovered" world, to define the contours of species being, nostalgia, and indeed history as such is empathy, rather than ontology. Rather than asking whether a being *is* or *is not* any particular type of creature, the novel implores us to ask whether a being can recognize another as *like itself* without recourse to its ontological status.

In his *Untimely Meditations*, Friedrich Nietzsche speculated on the relationship between temporality and animal being. He wrote that man

> also wonders at himself, that he cannot learn to forget but clings relentlessly to the past: however far and fast he may run, this chain runs with him. . . . Then the man says "I remember" and envies the animal, who at once forgets and for whom every moment really dies, sinks back into night and fog and extinguished forever. Thus the animal lives *unhistorically*: for it is contained in the present,

like a number without any awkward fraction left over. . . . That is why it affects him like a vision of a lost paradise to see the herds grazing. . . . If death at last brings the desired forgetting, by that act it at the same time extinguishes the present and all being and therewith sets the seal on the knowledge that being is only an uninterrupted has-been, a thing that lives by negating, consuming and contradicting itself.[1]

Regardless of whether we may know or not know about any given animal's experience of time, Nietzsche's provocation that animals live "unhistorically" acknowledges that temporality and a certain relation to the concept of history define at least a part of the discourse of species difference. But how do the stakes change once the past becomes eminently visible—namely, after an apocalypse? When the world ends, does the animal end with it? Cataclysm reduces both human and animal populations to this "awkward fraction" and makes the past all the more salient and irrevocable. What role can animals play in science fiction imaginings of this type of universe? In *Do Androids Dream*, not only do ersatz animals explicitly mark the apex of the animal's unequivocal entry into the commodity fetishism of modern and postmodern capitalism, but they also represent an attempt to impose a preapocalyptic structure on power relations. Just as the relationship between off-world humans and androids directly mimics slavery, the nostalgia of pet keeping functions as humanity's attempt to maintain the direct linearity of capitalist power structures. But, as I hope to demonstrate, one of the key insights of Dick's novel is to explore how the apocalypse makes possible new forms of species relations and that these new ethical relations are directly tied to the rupturing of history.

By thematizing animality in terms of an apocalyptic world, Dick offers a response to the question of what species relationships can look like after a radical rupture in the fabric of history. Specifically, the text proposes that the central problem of an apocalyptic world is a crisis of recognizability, a crisis marked by the instability of species binaries and the failure of the progressive, teleological myth of linear time. The act of recognizing acquires new meaning in a posthistorical world: to recognize ceases to be a question of materiality or ontology but instead becomes a relational question, of new paradigms of social, ethical, and kin relations. This chapter does not dissemble to be an overview of science fiction's various attempts to work through species relations. Instead, I want to take Dick's novel as the entry point for linking the critical deconstruction of history with the reframing of animality and species. By offering notes toward a transspecies eschatology, I want to sketch the contours of how a being that constitutes Nietzsche's "awkward fraction" can mobilize apocalyptic time to expand what it means to recognize a being as such. If, as I hope to demonstrate through the critical theory of Walter Benjamin and Giorgio Agamben, the form of the species boundary is historically contingent, then we can find that the apocalyptic end of history can present us with an uncovered set of problems about the ways in which humanity can relate to nonhuman animals and to its own animality: going beyond the historicity of human-animal relations allows for new forms of recognition, encounter, and ethics based on relational rather than ontological categories.

The plot of *Do Androids Dream* revolves around a nearly lifeless Earth decimated by "World War Terminus," a generic apocalypse of the nuclear variety. The majority of surviving humans have emigrated to Mars or other off-world colonies, where they are waited on by the androids, whose bondage is matched only by their flawless emulation of human physiology. Those humans who remain on Earth do so either out of nostalgia, stubbornness, or genetic corruption, victims to the ubiquitous dust falling like snow from the radioactive sky. The vast majority of nonhuman animal species are now extinct, and of the few remaining "authentic" species, only a handful of representatives remain. A virulent economy has developed around animals, to the point that human-animal relations have become distilled into a hypereconomic pet keeping. But to meet the demand for animal ownership, a separate industry specializes in the manufacture and upkeep of ersatz electric animals—complete with a monthly price guide informing its readers of the current prices of every desirable species (even those considered extinct, should one be fortunate enough to discover one). In this system, it is the institution, and the social capital it confers, that engenders Deckard's covetousness toward his neighbor's "real" mare: her fertility represents the possibility of economic gain, not an opportunity to recover the species from the brink of extinction. Like their humanoid counterparts, these electric animals are behaviorally and physiologically identical (at least externally) to their biological models. Inside, the electric animals contain electrical viscera; the opening of an "access panel" quickly dispels the illusion of the biological. But for the androids, material confirmation of their status as simulacra is generally performed postmortem and must delve to depths of bone marrow.

Figure 1. Jacket cover copyright © 1968 by Doubleday, a division of Random House, Inc., from *Do Androids Dream of Electric Sheep?* by Philip K. Dick. Used by permission of Doubleday, a division of Random House, Inc.

Instead, specialized police bounty hunters, of which Deckard is one, must conduct a test for empathy. This test uses physiology only as an indirect indication of concern for other living creatures by gauging the subject's response to scenarios involving animal death.

Do Androids Dream is Dick's most sustained exploration of the authenticity of human identity and experience, but, as Sherryl Vint has argued, "the importance of animals, electric and real" has been chronically "neglected in both the film adaptation and criticism."[2] Indeed, the majority of the scholarly attention devoted to *Blade Runner* centers principally around either postmodern subjectivity or cinematic and urban life. While the film highlights the dystopian and urban aspects of Dick's text and mostly elides his commentary on animals and animality, key associations between the vanished past, empathy, and the question of recognition remain salient. As Scott Bukatman argues in his seminal *Terminal Identity*, Ridley Scott's interpretation of Dick's vision describes "a future built upon the detritus of a retrofitted past (our present) in which the city exists as a spectacular site; a future in which the nostalgia for a simulacrum of history in the forms of the *film noir* (narrationally) and forties fashion (diegetically) dominates. . . . Most urgently, it is a future in which subjectivity and emotional affect are the signs of the nonhuman."[3] The simulacrum is the vehicle for a performance of nostalgia that is also a collective, sublimated mourning. But too much can be made of the presence of the animal and android simulacrum; *Do Androids Dream* is a text uncomfortably situated between the high modernist and postmodernist modes, and as such our interpretation must balance the Cold War specter of nuclear apocalypse against the technological threats to subjectivity and identity.

Although there are notable exceptions, the film generally avoids the novel's insistence on animals as the keystone in determining what counts as alive in postapocalyptic civilization. The humanoid cyborg occupies the thematic heart of the film, while in the novel androids provide only the narrative impetus. By focusing so directly on the human-android dyad, the film begs questions as to Deckard's status as an android—or, in the film's parlance, a "replicant," a term that foregrounds the cyborg as mimetic copy.[4] Maintaining a focus on the novel reveals this question to be moot. Deckard's status remains ambiguous because he comes to understand that empathy tests should include rather than exclude. Rather than a compartmentalizing binary, the androids function in a complex triangle of relations between human, cyborg, and animal. Dick takes animals as his central problem because they are the fundamental litmus test for the blurred boundaries in a postapocalyptic world. Humans have invested entire legal, political, and economic institutions to policing (and thereby reinforcing) the anxious border between human and android, using animals as the criterion while simultaneously relegating those animals to a hyperbolic commodity fetishism. The reverential and nostalgic economy centered on animals, both "real" and "ersatz," reifies preapocalyptic species relations. Moreover, the empathy questions—such as asking the subject to imagine receiving a calfskin wallet as a gift—do not distinguish whether the animal in question is biological or electric. The tests assume the animals in question are "real," not for the purposes of compassion for animal suffering but for the loss of a valuable commodity.

The connections between animals, technology, and nostalgia have already been elaborated both in animal studies and science fiction studies.[5] Building on John Berger's foundational dictum, we can note the advent of representations of and representational technologies for animals in correlation with the extirpation of animals and their enshrinement as monuments: "Everywhere animals disappear. In zoos they constitute the living monument to their own disappearance. . . . And in doing so, they provoked their last metaphor."[6] Dick's electric animals are the quintessential monument to the technological destruction of biological life as well as, in Sherryl Vint's argument, the embodiment of Deckard's alienation from his species being.[7] Akira Lippit claims that, in the context of technological modernity, "animals appeared to merge with the new technological bodies that were replacing them."[8] Here, they are identical. When Deckard considers his rooftop sheep, he is confronting the apotheosis of "modern technology . . . seen as a massive mourning apparatus, summoned to incorporate a disappearing animal presence that could not be properly mourned because . . . it was necessary to find a place to which animal being could be transferred, maintained in its distance from the world."[9] In the culmination of time and the possibility of an escape from temporal teleology, however, Deckard grasps the possibility of moving beyond technology-as-nostalgia and animal-as-metaphor. Animals have been maintained at the greatest possible distance from society at the same time as they are technologically and economically trapped in it. In the desert, Deckard exists, albeit temporarily, at a similar distance from the ruins of society. While Deckard's line of escape depends on the confluence of apocalyptic time with his empathetic response to artificial life, Deckard's transformation is a qualified one, mitigated by his initial profit motive and his begrudging acceptance of the electric toad into his family structure; his elation initially hinges on the anticipation of "a star of honor from the U.N. and a stipend.

A reward running into the millions of dollars."[10] What matters, however, is that Deckard has begun to slough off the weight of techno-speciesist discourse that had previously justified his state-sanctioned killing of androids.

Before continuing deeper into Dick's novel, I want to make clear the stakes of explicating apocalypse and animality as the guiding terms of this eschatology. While *Do Androids Dream* is literally apocalyptic, it also confronts the reader with a world radically stripped of its notions of history; the apocalypse gives lie to the myth of linear time and the notion of progressive civilization. Etymologically, the apocalypse brings together the Greek *apo* (to take off) with *kalyptein* (to cover) in order to describe a revelation, disclosure, or uncovering. By framing the reification and technologization of the human-animal-machine boundary in terms of the apocalypse, Dick forces the reader to consider what is uncovered about species relations, specifically the degree to which aspects of these relations are available for recontextualization. Deckard, after encountering the limits of empathy and linear temporality, encounters the full implications of this uncovering. In this context, when I discuss the historicity of the human/ animal relations, I mean that the end of history necessarily entails a rethinking of species relations. Starting with Foucault's dictum at the beginning of *The Order of Things*, we note that humanity as a discursive category emerges along with modernity as historically contingent. Thus Foucault writes that "what is available to archaeological analysis is . . . the threshold that separates us from Classical thought and constitutes our modernity. It was upon this threshold that the strange figure called man first appeared and revealed a space proper to the human sciences."[11] If we believe Foucault, we can begin to suspect that if the category of "man" is historically constructed, then the end of that history could mean the discursive end of "history."

Figure 2. Archontes with duck heads. One of the *abraxas* that prompted Agamben's reflections on the relationship between theriocephaly and the apocalypse. Images of the *abraxas* were originally included in Bataille's essay, "Le bas matérialisme et la Gnose," published in English as "Base Materialism and Gnosticism" in *Visions of Excess: Selected Writings, 1927–1939* (Minneapolis: University of Minnesota Press, 1985). (Source: Images copyright Bibliothèque nationale de France, Cabinet des Médailles, 2108 B; used with permission)

In any case, Foucault is not necessarily concerned with discursive structures after the apocalypse, and here we can diverge from him in two ways. First, instead of conducting an archaeological excavation of the history of humanity, we can search for an apocalyptic vision of a hypothetical, posthistorical humanity. Second, we can shift the deconstructive emphasis from the human sciences to the species boundary. That theorizing history should quickly dissolve into the question of the *end* of history for science fiction writers is hardly surprising given the twin specters of totalitarianism and atomic annihilation that haunted the art and philosophy of the twentieth century, and continue to haunt, in new biopolitical guises, the twenty-first century. But why should this potential for recognition and encounter across the species barrier take place in the aftermath of an apocalypse—that is to say, after the end of history? It might be objected that the apocalypse does not necessarily entail the "end" of history as such: indeed, it is precisely the vestigial elements of preapocalyptic institutions and discourses that Dick's novel exists to critique. I would respond that this is precisely the point of *Do Androids Dream*; it is only through Deckard's (albeit tentative) acceptance of the more radical ramification of the apocalypse—the seizing of different structures of time and categories of empathy—that the break from the past becomes complete and productive.

In order to connect the apocalypse to questions of species, we must return to Nietzsche's awkward fraction: what remains of man, and of his relationship to his own animality? It is exactly this question that motivates Giorgio Agamben at the opening of his *The Open: Man and Animal*. This shift in emphasis—from reconstructed past to imagined future, from man as such to man in the context of animality—aligns us with Agamben's explication of the interpenetration of the human/animal relationship with the theorization of history. Agamben finds inspiration in illustrations from a thirteenth-century Hebrew Bible depicting figures at a messianic banquet with animal heads: an eagle, an ox, a lion, an ass, a leopard, and a monkey. What are we to make of this spiritual theriocephaly? This question, and the consternation it caused Bataille, prompts Agamben to argue that it represents the conclusion of humanity and the end of the telos of messianic—and, Agamben would add, Hegelian and Marxist—history. Hardly a benign question, it not only asks what remains of man and his animality after the end of history, but it also suggests that "on the last day, the relations between animals and men will take on a new form, and that man himself will be reconciled with his animal nature."[12] While such reconciliation is by no means given—Dick's text clearly argues that achieving it is a difficult if not Sisyphean task—it becomes clear that Agamben identifies the transition from historical to posthistorical as a critical moment, a moment in which species distinctions are eligible for radical reconsideration, a moment that we can identify as postapocalyptic. Agamben writes:

> The messianic end of history of the completion of the divine *oikonomia* of salvation defines a critical threshold, at which the difference between animal and human, which is so decisive for our culture, threatens to vanish. That is to say, the relation between man and animals marks the boundary of essential domain, in which historical inquiry must necessarily confront that fringe of ultrahistory which cannot be reached without recourse to first philosophy. . . . If animal life and human life could be superimposed perfectly, then neither man nor animal—and, perhaps, not even the divine—would any longer be thinkable. For this reason, the arrival at posthistory necessarily entails the reactualization of the prehistoric threshold at which that border had been defined. Paradise calls Eden back into question.[13]

It is clear that, for Agamben, the end of history halts the dialectical process—which he later refers to as the "anthropological machine"—in which man becomes defined only as that creature that must recognize itself in opposition to the animal in order to exist. The apocalyptic end of history is also the radical superimposition of the human and its other, which engenders a profound sense of confusion and anxiety for a world in which humanity is only defined as that which is *not* other. The final point of this passage is that the end of history entails, in its completion of the telos of the human, a distillation of identity and, most pressingly, a reduction in number. As the reference to Eden indicates, paradise (which is both before and after history) comprises only the blessed, or perhaps the cursed. For the meantime, the question becomes not merely who survives, but what survives. The presence of nonhuman animality in depictions of postapocalypse upsets the superiority of humanity. This explains not only the use of animals as the test case for delineating human from android but also the institutional boundary work to segregate "specials" and "chickenheads"—again, the metaphor of theriocephaly—from those individuals fit to reproduce and emigrate.

What truly perplexed Agamben is that these figures *are* blessed in that they are open to new permutations of the human-animal relationship: is this what remains of "man" at the end of history, and if so, how do we interpret the animality we see there? It is not the end of animality as such, but rather the mysterious hybridization of human and animal. The question of the remnant, the awkward fraction, guides Agamben's query as much as Nietzsche's. Nietzsche, however, situates the animal in an unhistorical present, whereas Agamben interrogates the presence of animal bodies outside history. *Do Androids Dream* accomplishes both of these tasks, because Dick's bodies refer both to a material, contaminated present and a vanished past: the electric animals inevitably refer to the near absence of biological animals, and the human bodies bear the contaminating trace of the dust. This recalls Etienne Benson's suggestion, in his contribution to this volume, that the animal trace is always in the past, always diachronic, always referencing a before and after. Species being in this world is always marked by being left over from historical rupture, of being carried over into an ambiguous, always-disappearing time. This is what Nietzsche means when he claims that death reveals the fallacy of continuous being and heterogeneous time. To further clarify this, it is useful to note the difference between

Figure 3. Acephalic god beneath two animal heads. At the foot of the god, in a circle formed by a serpent biting its tail, Anubis, a woman, and a dog; below, a mummy. The acephalic god can be identified with the Egyptian god Bes. (Source: *Abraxas* image copyright Bibliothèque nationale de France, Cabinet des Médailles, 2170; used with permission)

chronos—indicating chronological, sequential, and quantitative time—and *kairos*—which is qualitative and refers roughly to the "proper moment," or the moment of opportunity. *Kairos* describes time outside of or in opposition to the time of the sovereign, a messianic time brought about here by the apocalypse.[14] The confluence of encounters with nonchronological time and with postapocalyptic animality brings Deckard to his final realization. Apocalypse is the essence of *kairos* for species relations, and the survivors of the catastrophe, regardless of their species or ontological status, find themselves presented with an immense opportunity.[15] Because apocalypse fundamentally disrupts the temporal structures on which humanity defines itself in the mirror of animality, the *kairos* at the end of history is a chance to seize new relations of ethics, kinship, and life.

It is at this point that Deckard's final animal encounter becomes the fulcrum on which this reading of the novel hinges. Deckard, although he has succeeded in dispatching (or "retiring") the renegade androids charged to him, falls into despair and flees the decimated San Francisco for the expanses of desert between the cities, landscapes that may have once contained suburbs or rural farms. There, he undergoes a quasi-mystical fusing with the figure of Wilbur Mercer, a spiritual merging with the body of the singular Mercer as well as with all other adherents around the world similarly engaged in the cult of "Mercerism." For once, Deckard experiences empathy not as a test but rather than as overwhelming and penetrating fact, an empathetic merging that dissolves the boundaries of the individual rather than reinforcing them. Following this experience, Deckard, depleted, is about to leave the desert when he catches sight of an animal almost invisible in the barren soil. Betrayed by its feeble movement, this animal is a toad, one that Deckard believes to be real. Ecstatic, Deckard rescues the toad and brings it home, where his wife, Iran, promptly discovers an access panel and exposes the toad's electric nature. Despite his initial disappointment, Deckard eventually decides to welcome the toad into their family structure, transcending his earlier obsession with obtaining a real animal to compensate for the embarrassment of his electric sheep. He concludes that "the electric things have their lives, too. Paltry as those lives are."[16] The logic of examination and identification

Figure 4. God with the legs of a man, the body of a serpent, and the head of a cock. (Source: *Abraxas* image copyright Bibliothèque nationale de France, Cabinet des Médailles, M. 8003; used with permission)

had previously defined human attitudes toward artificial life. However, Deckard's encounter marks a difference based not on a rational taxonomy but on an opening of categories of empathetic concern.

Recognizing this empathetic opening, and its dependence on animality, alters the stakes of interpreting *Do Androids Dream*, a stance taken by both Ursula Heise and Sherryl Vint. In terms of ecological criticism and speciesist discourse, respectively, these authors have set out essential groundwork for incorporating animality into discussions of Dick's novel. In summarizing the recent decades of science fiction representation and cyborg discourse in scholarship, Heise argues that "sometimes implicitly, and quite often explicitly, the extinction of real animal species crucially shapes the way in which . . . artificial animal forms are approached and evaluated."[17] This fundamental assertion of the influence of extinction on the representation of artificial animals contains within it the logic of apocalypse, but Heise is perhaps less sanguine about Deckard's transformation than I. She reads Deckard's toad as an "imaginative gesture" that signifies "the recuperation by humans of lost animal species," arguing that "this recuperation is mediated and in the end contained by advanced technology."[18] I agree that technology continues to define Deckard's encounter, but I would also add that it is the uncovering logic of the apocalypse that allows Deckard to see beyond the simulacral skin, beneath the electrical pelt. Heise points out that "even though electric animals are common in Dick's world . . . their artificiality is carefully concealed from the neighbors."[19] Moreover, an asymmetry seems to persist, insofar as Deckard's empathetic response to the toad will not necessarily extend to androids.[20] However, Heise nuances her reading of Dick by suggesting that Deckard's encounter, while not expressing a fully realized or a "post-Haraway perspective," nonetheless allows the reader to push past the speciesism that otherwise pervades *Do Androids Dream*. This is precisely the argument of Vint's article, which maintains that Deckard's narrative is one of recuperating, in the Marxist sense, his species being: "So long as they remain trapped in a world dominated by commodity logic, their lives will remain tainted by something of [commodity logic]. . . . At the same time, however, the care offered to the electric toad also exceeds this logic and embraces something of human species being, a connection that is not about exchanging care for value, a relationship different than that Deckard had with his electric sheep."[21]

From this perspective, Deckard is in a unique position to recognize similarities between human, animal, and artificial life (either android or animal) and thus to provide the outline of "the advocacy of the cyborg animal," which is "at least in part a call to abandon speciesist prejudice and accept alternative life-forms as beings with an existence and rights of their own."[22] While this is a relatively small gesture, it represents a radical reinterpretation of the novel's central problem, namely, the one of recognition. As Vint points out, Dick wrote that he considers the difference between so-called authentic biological life (humans, animals) and ersatz life (androids, electric animals) as applying "not to origin or to any ontology but to a way of being in the world."[23] This quote is essential because it frames the eschatological questions of *Do Androids Dream* as a matter of praxis rather than essence. Although Dick seems to prefer the term "distinguish," I opt for "recognize" because it moves us toward incorporation rather than segregation. The android empathy test—or rather, the putative human empathy it takes as its benchmark—privileges not empathy per se, but empathy only for correct objects (wanting to protect animals from death) while condemning a lack of affect for bad objects (revulsion at the calfskin wallet). Dick recognizes the hypocrisy of using such revulsion to sift android from human: even though androids do not recoil in horror from the explicitly commodified dead animal, humans are able to coexist in their malaise with similarly commodified live animals. Indeed, human empathy itself has become ersatz, as Dick demonstrates through the pervasive use of "mood organs," from which humans dial specific emotions: 481 for an "awareness of the manifold possibilities open to me in the future," or 888 for "the desire to watch TV no matter what's on."

While the androids fail to perfectly mime human repulsion to animal death, the Pavlovian humans can only experience simulated emotion. This is why the novel's conception of the New Age religion Mercerism, in which human supplicants experience a virtual "fusing" with the Sisyphean figure of Mercer, represents Deckard's way out. In this experience, individuals become aware of the emotions and experiences of all other humans as the frail, elderly Mercer scrambles up a rocky mountain slope. The distinction between mood organs and Mercerism is that the former isolates while the latter collapses individual identity into a paradoxical collective singularity. Even when the androids expose Mercerism as fabricated and mediated—merely a recycled Hollywood film clip from before the war—this makes little difference: what matters is the connection between individuals or, more specifically, recognizing that empathy is not a matter of maintaining the stability of identity but of a radical capacity for fusing with other creatures. This is what Heise means when she concludes that "the animal cyborg can take us, through *the discovery of otherness in our own technological creations*, to the recognition of and respect for the nonhuman others we did not make."[24]

But beyond expanding the scope of empathetic concern, Deckard's experience with Mercerism makes salient the altered relationship to history. An essential component of Mercer's cosmology is that Mercer, having failed to reach the never-seen summit of the mountain, tumbles into a deep pit filled with the skeletons of all the animals eradicated by the apocalypse. After the passage of an indeterminately long period of time in this graveyard, Mercer is able to bring these skeletons back to life, and via their resurrection he is lifted out of the pit. The cycle repeats. While it is possible to interpret this scene as another reassertion of humanity's desire to absolve itself of the guilt of extinction, we cannot overlook the import of the near-miraculous reanimation: it is through this ecumenical revivification of all species that Mercer is able to distinguish his version of collective identification from the exclusionary empathy Deckard practices as a bounty hunter. Similarly, it is only through the animal technology that Deckard becomes open to considering the "paltry life" of electric things. This animation of animal corpses also reveals the cyclical nature of Mercerian time, a schema that Deckard can only see after the end of traditional history. This vision, as well as the glimpse of the solitary toad, affixes to Deckard and ruptures his vision of the world like Barthes's *punctum*: "In this glum desert, suddenly a specific photograph reaches me; it animates me, and I animate it. So that is how I must name the attraction which makes it exist: an *animation*."[25]

Here, there is a reciprocal animation between Mercer and the animal skeletons. They rise together, and reading this scene as an escape—from linear chronology as well as from death—allows us to pinpoint why it should be the electric animals that provide Deckard a new form of empathy. To understand this, it is useful to quote Walter Benjamin's description of precisely this scenario: "What are phenomena rescued from? Not only, and not in the main, from the discredit and neglect into which they have fallen, but from the catastrophe represented very often by a certain strain in their dissemination, their 'enshrinement as heritage'—They are saved through the exhibition of the fissure within them—there is a tradition that is catastrophe."[26]

In the economy of apocalyptic commodity fetishism, the animal is an object of nostalgia, chaining humanity to its traumatic past and defining its alienated present. But, as Benjamin describes, the "rescue" from the monumentality of animals—their "enshrinement as heritage"—occurs via the exposure of an internal rift: in *Do Android Dream*, this internal fissure resides in the gap between electric viscera and biological facade. Electric animals make this fissure explicit, while androids hide it. Mercer's circular timeline exposes in a moment of timeless purgatory the seam where a previously linear chronology wraps around, connecting its end to its beginning. But we must not consider the exposure of this fissure to be a weakness. Dick's insight is to thematize this internal fissure—what Agamben calls an "intimate caesura"—and then undercut its material embodiment in the muddled physiological distinction between human and android, "real" and electric animal. With Deckard's animal encounter, Dick not only avoids ontology, but his cyclical cosmology obviates traditional conceptions of chronology as well. Moreover, the body of the electric animal is the site of the interpenetration of historical and technological forces, and thus the vehicle by which these themes are made visible. As Benjamin writes: it is "in the form of the historical confrontation that makes up the interior (and, as it were, the bowels) of the historical object, and into which all the forces and interests of history enter on a reduced scale."[27] Deckard's triumph, provisional though it might be, is to transform the nature of this confrontation from exclusionary violence to inclusive empathy. In the process, he recognizes the electric animal in its historicity—that is to say, the degree to which nostalgia, economy, speciesism, and apocalypse coalesce in the animal body—and attempts to extricate it from its strictures, just as Mercer reanimates life and time. This move eschews the debate between real and simulated in order to think an ethics of recognition rather than an ethics of materiality or ontology. Deckard's toad was nearly invisible against the desert terrain, and this camouflage defines the problem of visualizing and encountering life in a shattered world. Dick writes: "So this is what Mercer sees . . . life which we can no longer distinguish; life carefully buried up to its forehead in the carcass of a dead world."[28]

In *Do Androids Dream*, the historical contingency of species relations comes into direct relation with nostalgic technologies of animal mourning. Just as the apocalypse uncovers, for Agamben and Benjamin, the possibility of cyclical or messianic time, Heidegger understands that "the essence of technology lies in Enframing."[29] The "Enframing" qualities of technology go beyond the mechanico-electrical endoskeleton and biological exoskeleton; instead, the destination of the "progress" of technology reveals the inherent danger in both the teleology of technology and its simultaneous temporal meanings. For Heidegger, the destiny or destination marks the danger of technology's revealing-as-enframing: "The destining of revealing is as such, in every one of its modes, *danger* . . . not just any danger, but danger as such."[30] Among the possible ways to gloss "danger as such" we can choose the apocalypse not only for its similar properties of revealing but also for the technological means of its coming-to-be. As James Leo Cahill has noted in his comments on the etymology of "apocalypse," Heidegger "argues that Enframed revealing—a mode of calculation—is potentially world-ending, in a sense, apocalyptic."[31] But, as Heise has noted in her discussion, the specter of the nuclear apocalypse remains background noise in Deckard's world, an unassailable fact that nonetheless remains ambiguous. In this sense, *Do Androids Dream* is resolutely postapocalyptic, a trait that helps obviate the eventual technological anachronisms that often plague science fiction. This ambiguity also cuts to the quick of Heidegger's characterization of technological danger: "The threat to man does not come in the first instance from the potentially lethal machines and apparatus of technology. . . . The actual threat has already affected man in his essence. . . . The rule of enframing threatens man with the possibility that it could be denied to him to enter into a more original revealing hence to experience the call of a more primal truth."[32]

In Deckard's world, the combination of apocalypse and technology has multiply penetrated both human and animal bodies. Recollecting Benjamin's argument concerning the role of the internal fissure in making visible the historicity of objects, we can understand the internalization of technological and species anxiety that allows Heidegger to warn that the "threat has already affected man in his essence."[33] This is because, as Foucault and Agamben both argue, the discursive separation of human and animal already exists within the category of "man": "The division of life into vegetal and relational, organic and animal, animal and human, therefore passes first of all *as a mobile border within living man*, and without this intimate caesura the very decision of what is human and what is not would probably not exist."[34] In the postapocalyptic world of *Do Androids Dream*, animal technology has allowed humans to project this "intimate caesura" onto and into the body of the artificial other.

Deckard's sheep resides on the roof, head down in its meager skyward-facing pasture, subject to the slowly falling atomic dust. Unlike William Gibson's technologically opaque sky "the color of television tuned to a dead channel,"[35] Dick's nuclear sky obscures by emitting a different type of static: the mutating fallout that transforms humanity into a "special" class of subaltern at the same time as it relegates biological animals to valued relics and electric animals to nostalgic commodity. The biological is the ruin, the electric the monument. Deckard and his neighbors keep their animals away from the earth, exactly opposed to the skeletal pit from which Mercer arises on the back of reanimated animals. The animal chained to the roof is, as Nietzsche said of the human, also chained to the past; this animal-as-nostalgia resides near the sky, near to the polluting rain of history. Mercer's animals, the animal-as-encounter, emerge from the earth to allow entry to a cyclical time. Somewhere between Mercer's pit and the atomic sky, Deckard is able to encounter the toad that frees him from "the tyranny of an object."[36] In the desert, the liminal space between the decaying cities, Deckard begins to transcend the hierarchy implied by preapocalyptic time. As Vint points out, this embrace of a new paradigm of species relations is Dick's contribution to the critique of Cartesian subjectivity: "although Deckard is supposed to rationalize his work as a bounty hunter while still theoretically maintaining his reverence for empathy, he comes to realize that doing so requires precisely the sort of affect and cognition split that makes him both a proper Cartesian subject yet also an android."[37] Here we see the suggestion that *Do Androids Dream* expresses the possibility of a realignment of subjectivity along new lines of recognition for objects previously excluded from ethical consideration.

Nowhere in the novel is this realignment more profound than with the example of Mercer. As with the bodily involvement of the Mercerian supplicant, it is difficult to untangle the materiality of the collective experience. When stones strike Mercer's body, the body of the individual participant can also be wounded. But, if we are to take seriously the novel's imperative, as expressed through Mercer, to abandon the biological-ersatz distinction as the basis for inclusion in an empathetic sphere, then we must read Mercer's resurrection of the animals differently. The reanimation of the corpses can, on the one hand, satisfy the same impulse toward nostalgia and restoration inherent in rooftop pet keeping: the fantasy of undoing the mass extinction allows for an obviation of guilt and the continuation of a Sisyphean struggle against decay and entropy. But the other lesson of Mercer's resurrection, the meaning that most directly embraces the meaning of apocalyptic time, is one of reanimation: the bringing to life of things actively stripped by anthropocentric discourse of their subjectivity. Recognizing the potential for life in the mountain of bones mirrors the possibility of empathizing with a being regardless of its putative status as real or ersatz, human or android. When Deckard learns to care for the toad he initially believed to be real, he, too, glimpses the postapocalyptic capacity for escaping the tyranny of objects. This tyranny is reciprocal—androids must fear "retirement" at the hands of bounty hunters, while humans dread the unseating of their tenuous grasp on the coherence of the past and their own identities—and manifests in the hatred Deckard initially feels toward his sheep.

If, as I have argued, the rupturing of time inherent in the apocalypse implies an uncovering of species relations, then we can read Dick's narrative as culminating in the seizing of a "proper moment" of encounter. *Do Androids Dream* filters species and temporality through the celestial and the earthly, the oneiric and the material. Where *Blade Runner* emphasizes the visuality of the encounters between human, android, and animal, *Do Androids Dream* concerns itself with the potential hollowness of the material, the possibility of sloughing off the bounds of the material like so much snakeskin.

NOTES

1. Friedrich Nietzsche, *Untimely Meditations*, ed. Daniel Breazeale, trans. R. J. Hollingdale (Cambridge: Cambridge University Press, 1997), 61.
2. Sherryl Vint, "Speciesism and Species Being in *Do Androids Dream of Electric Sheep?*" *Mosaic* 40, no. 1 (2007): 111.
3. Scott Bukatman, *Terminal Identity: The Virtual Subject in Postmodern Science Fiction* (Durham, N.C.: Duke University Press, 1993), 131.
4. I would suggest that the term "android," by combining the Greek *andro* (man) with *oid* (resembling, related to *eidolon*), references machines that have been cloaked in the skin of biological life. Indeed, the *Oxford English Dictionary* concisely defines android as "an automaton resembling a human being." Whereas an android may be a mechanical or electrical framework covered with or shaped as living tissue (consider the scene from James Cameron's *Terminator 2* in which Arnold Schwarzenegger's character cuts away the flesh of his forearm to reveal his steel endoskeleton), the replicant may be materially indistinguishable from the thing it is meant to copy or mimic. This explains the focus on access panels for the ersatz animals, which characters often open in order to reveal the wiring inside; indeed, the revelation of the final toad as electric occurs when Deckard's wife opens such a panel. The humanoid cyborgs do not seem to have these panels.
5. For an overview of the relationship between science fiction and animal studies, see Sherryl Vint, "'The Animals in That Country': Science Fiction and Animal Studies," *Science Fiction Studies* 105 (Fall 2008), 177–178; http://www.depauw.edu/sfs/abstracts/a105.htm#vint.
6. John Berger, "Why Look at Animals?" in *About Looking* (New York: Vintage International, 1991), 26.
7. Vint, *Speciesism and Species Being*, 124–125.
8. Akira Mizuta Lippit, ". . . From Wild Technology to Electric Animal," in *Representing Animals*, ed. Nigel Rothfels (Bloomington: Indiana University Press, 2002), 124.
9. Lippit, "From Wild Technology," 125–126.
10. Philip K. Dick, *Do Androids Dream of Electric Sheep?* (New York: Del Rey, 1996), 237.
11. Michel Foucault, *The Order of Things* (New York: Vintage, 1991), xxvi.
12. Giorgio Agamben, *The Open: Man and Animal*, trans. Kevin Attell (Stanford, Calif.: Stanford University Press, 2004), 3.

13. Agamben, *Open*, 21.

14. Agamben fully elaborates his notion of *kairos* in *The Time That Remains: A Commentary on the Letter to the Romans*, trans. Patricia Dailey (Palo Alto, Calif.: Stanford University Press, 2005).

15. Benjamin defines catastrophe as "having missed the opportunity." Walter Benjamin, *The Arcades Project*, trans. Howard Eiland and Kevin McLaughlin (Cambridge, Mass.: Belknap Press, 1999), section N10,2, 474.

16. Dick, *Do Androids Dream?* 241.

17. Ursula K. Heise, "From Extinction to Electronics: Dead Frogs, Live Dinosaurs, and Electric Sheep," in *Zoontologies: The Question of the Animal*, ed. Cary Wolfe (Minneapolis: University of Minnesota Press, 2003), 60.

18. Heise, "Extinction to Electronics," 72.

19. Heise, "Extinction to Electronics," 72.

20. Fully resolving this issue would require a gloss of Deckard's relationship with Rachel, especially their sexual encounter prior to Deckard's final assault on the renegade androids. How one reads the ambiguity of his attitude toward loving the machine, which translates to anxieties around transspecies sex, would influence one's evaluation of Deckard's future relationship to androids.

21. Vint, "Speciesism and Species Being," 124.

22. Heise, "Extinction to Electronics," 77.

23. Philip K. Dick, "Man, Android and Machine," in *The Shifting Realities of Philip K. Dick: Selected Literary and Philosophical Writings*, ed. Lawrence Sutin (New York: Vintage, 1996), 212, quoted in Vint, "Speciesism and Species Being," 125.

24. Heise, "From Extinction to Electronics," 78, emphasis added.

25. Roland Barthes, *Camera Lucida* (New York: Hill and Wang, 1982), 20. Akira Lippit uses this quote as the epigraph to his section on "Cinema" in "From Wild Technology to Electric Animal."

26. Benjamin, *Arcades Project*, section N9,4, 473.

27. Benjamin, *Arcades Project*, section N10,3, 475.

28. Dick, *Do Androids Dream?* 236.

29. Martin Heidegger, *The Question Concerning Technology*, trans. William Lovitt (New York: Garland, 1977), 26.

30. Heidegger, *Question Concerning Technology*, 26.

31. James Leo Cahill, "Anacinema: Peter Tscherkassky's Cinematic Breakdowns: Towards the Unspeakable Film," *Spectator* 28, no. 2 (Fall 2008): 101n.36.

32. Heidegger, *Question Concerning Technology*, 28.

33. Heidegger, *Question Concerning Technology*, 28.

34. Agamben, *Open*, 15, emphasis added.

35. William Gibson, *Neuromancer* (New York: Ace Trade, 2000), 3.

36. Dick, *Do Androids Dream?* 42.

37. Vint, "Speciesism and Species Being," 117.

Bibliography

Abram, David. 1996. *The Spell of the Sensuous: Perception and Language in a More-Than-Human World*. New York: Vintage Books.

Adelard of Bath. 1998. *Conversations with His Nephew: On the Same and the Different, Questions on Natural Science and On Birds*. Translated and edited by Charles Burnett. Cambridge: Cambridge University Press.

Agamben, Giorgio. 2003. *The Open: Man and Animal*. Translated by Kevin Attell. Palo Alto, Calif.: Stanford University Press.

Agamben, Giorgio. 2005. *The Time That Remains: A Commentary on the Letter to the Romans*. Translated by Patricia Dailey. Palo Alto, Calif.: Stanford University Press.

Alberti, Leon Battista. 1847. *Il Cane*. Translated by Piero di Marco Parenti Fiorentino. Ancona: Tipografia di Aurelj G. e comp.

Albertus Magnus. 1999. *On Animals. A Medieval Summa Zoologica*, 2 vols. Translated and edited by Kenneth F. Kitchell Jr. and Irven Michael Resnick. Baltimore: Johns Hopkins University Press.

American Kennel Club. 2006. *The Complete Dog Book*, 20th ed. New York: Ballantine Books.

American Kennel Club. n.d. Press release, "AKC Congratulates Obamas and Welcomes 'Bo,' Portuguese Water Dog to the White House." http://www.akc.org/poll/special/presidential.cfm.

Anderson, Virginia DeJohn. 2004. *Creatures of Empire: How Domestic Animals Transformed Early America*. New York: Oxford University Press.

Anker, Peder. 2001. *Imperial Ecology: Environmental Order in the British Empire, 1895–1945*. Cambridge, Mass.: Harvard University Press.

Anon. 1896. "Interview with Mr. Lionel Decle, the African Explorer: Interesting Facts about the Matabele." *African Review* (May 9): 917.

Anon. 1896. "The 'Rinderpest': What It Is with Symptoms and Causes: An Interview with Dr. Hutcheon." *African Review* (May 23): 1027.

Anon. 1896. "The Cattle Plague in South Africa: A Very Serious Outlook." *African Review* (June 6), n.p.

Anon. 1896. "The Hon. John Scott Montagu, MP, on the Matabeli Rising and Its Causes." *African Review* (June 6): 1121.

Anon. 1896. "The Situation in South Africa: Origin of the Matabeli Revolt." *African Review* (April 18): 752.

Anon. 1931. "Igloo, Byrd's Dog, Buried." *New York Times*, June 1.

Anon. 1931. "Igloo, Byrd's Dog, Polar Hero, Is Dead." *New York Times*, April 22.

Anon. 1932. "Rin Tin Tin Dies at 14 on Eve of 'Comeback.'" *New York Times*, August 11.

Anon. 1934. "Combating Foot and Mouth Disease—Drastic Slaughter Policy Adopted." *African World* (August 18): 181.

Anon. 1934. "Inter-State Veterinary Conference Proposed—to Investigate Foot-and-Mouth Question." *African World* (November 3): 101.

Anon. 2002. "*Denver Post, Rocky Mountain News* to Charge for Obituary Notices." *Denver Post*, March 24.

Anon. 2003. *Honoring Your Loved One in Print: Your Guide for Writing and Submitting a Personalized Obituary*. Champaign, Ill.: *News-Gazette*.

Anon. 2008. "Obama's First Press Conference Turns to Dog Talk: 'This Is a Major Issue.'" Associated Press/Huffington Post, November 7. http://www.huffingtonpost.com/2008/11/07/obamas-first-press-confer_n_142201.html.

Anon. 2009. "Is Bo a Rescue Pooch? Questions Linger Over Adoption of Obama Dog." Associated Press, April 13. http://www.foxnews.com/politics/2009/04/13/bo-rescue-pooch-questions-linger-adoption-obama-dog.

Anon. n.d. Battle at Kruger. http://www.youtube.com/watch?v=LU8DDYz68kM.

Aquinas, Saint Thomas. 1975. "Homicide." In *Summa Theologiae: Latin Text and English Translation, Introductions, Notes, Appendices and Glossaries*. Blackfriars edition, Vol. 38, 19–21. New York: McGraw-Hill.

Arbel, Benjamin. 2010. "The Attitude of Muslims to Animals: Renaissance Perceptions and Beyond." In *Animals and People in the Ottoman Empire*, edited by Suraiya Faroqhi. Istanbul: Muhittin Eren.

Arens, William. 1979. *The Man-Eating Myth: Anthropology and Anthropophagy*. New York: Oxford University Press.

Armstrong, Philip. 2008. *What Animals Mean in the Fiction of Modernity*. New York: Routledge.

Attenborough, David. 2003. *The Life of Mammals*. BBC Video. DVD.

Audubon, John James, and John Bachman. 1854. *The Quadrupeds of North America*, Vol. 3. New York: Audubon.

Augustine, Saint. 1995. *The City of God*. In *A Select Library of the Christian Church: Nicene and Post-Nicene Fathers*, Vol. 2, edited by Philip Scharff. Peabody, Mass.: Hendrickson.

Baker, Steve. 2001. *Picturing the Beast: Animals, Identity, and Representation*. Urbana: University of Illinois Press (also published in 1993).

Baker, Steve. 2002. "What Does Becoming-Animal Look Like?" In *Representing Animals*, edited by Nigel Rothfels, 67–98. Bloomington: Indiana University Press.

Baker, Steve. 2007. "What Is the Postmodern Animal?" In *The Animals Reader: The Essential Classic and Contemporary Writings*, edited by Linda Kalof and Amy Fitzgerald, 278–288. Oxford, U.K.: Berg.

Banting, Pamela. 2009. "Magic Is Afoot: Hoof Marks, Paw Prints and the Problem of Writing Wildly." In *Animal Encounters*, edited by Tom Tyler and Manuela Rossini, 27–44. Boston: Brill.

Baraz, Daniel. 1998. "Seneca, Ethics and the Body: The Treatment of Cruelty in Medieval Thought." *Journal of the History of Ideas* 59, no. 2: 195–215.

Barber, Malcolm. 2000. *The Cathars: Dualist Heretics in Languedoc in the High Middle Ages*. Harlow, U.K.: Longman.

Barrow, Mark V. 1998. *A Passion for Birds: American Ornithology after Audubon*. Princeton, N.J.: Princeton University Press.

Barthes, Roland. 1982. *Camera Lucida*. New York: Hill and Wang.

Bauman, Zygmunt. 2000. *Liquid Modernity*. Cambridge: Cambridge University Press.

Bayle, Pierre. 1965. *Historical and Critical Dictionary: Selections*. Translated and edited by Richard H. Popkin. Indianapolis: Bobbs-Merrill.

Beinart, William, and Peter Coates. 1995. *Environment and History: The Taming of Nature in the USA and South Africa*. New York: Routledge.

Beinart, William, Karen Brown, and Daniel Gilfoyle. 2009. "Experts and Expertise in Colonial Africa Reassessed: Colonial Science and the Interpenetration of Knowledge." *African Affairs* 108, no. 432: 413–433.

Benjamin, Walter. 1999. *The Arcades Project*. Translated by Howard Eiland and Kevin McLaughlin. Cambridge, Mass.: Belknap Press.

Benson, Elizabeth P., ed. 1970. *The Cult of the Feline*. Washington, D.C.: Dumbarton Oaks Research Library and Collections.

Benson, Elizabeth P. 1998. "The Lord, the Ruler: Jaguar Symbolism in the Americas." In *Icons of Power: Feline Symbolism in the Americas*, edited by Nicholas Saunders, 53–76. New York: Routledge.

Berger, John. 1991. "Why Look at Animals?" In *About Looking*, 3–28. New York: Vintage International.

Bernat Vistarini, Antonio, Emilio Bianco, John T. Cull, and Tomás Sajó. 2004. "His Master's Voice: Johannes Sambucus and His Dog Bombo." *Silva. Digital Review of Studiolum* 3. http://www.studiolum.com/en/silva3.htm.

Bishop, Morris. 1964. *Petrarch and His World*. London: Chatto & Windus.

Bland, K. P. 1979. "Tom-cat Odour and Other Pheromones in Feline Reproduction." *Veterinary Research Communications* 3, no. 1: 125–136.

Blasini, Gilberto. "The Arsenio Hall Show." The Museum of Broadcast Communications Web site. http://www.museum.tv/archives/etv/A/htmlA/arseniohall/arseniohall.htm.

Bloch, Marc. 1953. *The Historian's Craft*. New York: Knopf.

Bloor, David. 1976. *Knowledge and Social Imagery*. London: Routledge.

Blount, Margaret. 1974. *Animal Land*. London: Hutchinson.

Bonilla-Silva, Eduardo. 2002. "The Linguistics of Color-Blind Racism: How to Talk Nasty about Blacks without Sounding 'Racist.'" *Critical Sociology* 28, nos. 1–2: 41–64.

Boomgaard, Peter. 2001. *Frontiers of Fear: Tigers and People in the Malay World, 1600–1950*. New Haven, Conn.: Yale University Press.

Bowker, Geoffrey C. 2005. *Memory Practices in the Sciences*. Cambridge, Mass.: MIT Press.

Brody, Hugh. 2000. *The Other Side of Eden: Hunters, Farmers, and the Shaping of the World*. New York: North Point.

Brower, Matthew Francis. 2005. "Animal Traces: Early North American Wildlife Photography." Ph.D. diss., University of Rochester.

Brown, David, and Carlos Lopez Gonzalez. 2001. *Borderland Jaguars*. Salt Lake City: University of Utah Press.

Brown, Karen. 2001. "The Conservation and Utilization of the Natural World: Silviculture in the Cape Colony circa 1902–1910." *Environment and History* 7, no. 4: 427–447.

Brown, Karen. 2002. "Cultural Constructions of the Wild: The Rhetoric and Practice of Wildlife Conservation in the Cape Colony at the Turn of the Twentieth Century." *South African Historical Journal* 47:75–95.

Brown, Karen. 2003. "Agriculture in the Natural World: Progressivism, Conservation and the State. The Case of the Cape Colony in the Late 19th and Early 20th Centuries." *Kronos Special Edition on Environmental History* 29:109–138.

Brown, Karen. 2003. "Political Entomology: The Insectile Challenge to Agricultural Development in the Cape Colony 1895–1910." *Journal of Southern African Studies* 29, no. 2: 529–549.

Brown, Karen. 2003. "Trees, Forests and Communities: Some Historiographical Approaches to Environmental History on Africa." *Area* 35, no. 4: 343–356.

Brown, Karen. 2005. "Tropical Medicine and Animal Diseases: Onderstepoort and the Development of Veterinary Science in South Africa 1908–1950." *Journal of Southern African Studies* 31, no. 3: 513–529.

Brown, Karen. 2007. "Poisonous Plants, Pastoral Knowledge and Perceptions of Environmental Change in South Africa, c. 1880–1940." *Environment and History* 13, no. 3: 307–332.

Brown, Karen. 2008. "From Ubombo to Mkhuzi: Disease, Colonial Science and the Control of *Nagana* (Livestock Trypanosomiasis) in Zululand, South Africa, c. 1894–1955." *Journal of the History of Medicine and Allied Sciences* 63, no. 3: 285–322.

Brown, Karen. 2008. "Frontiers of Disease: Human Desire and Environmental Realities in the Rearing of Horses in 19th and 20th Century South Africa." *African Historical Review* 40, no. 1: 30–57.

Brown, Karen and Daniel Gilfoyle, eds. 2007. "Introduction: Livestock Diseases and Veterinary Science in Southern Africa." *South African Historical Journal* (Special Edition on Environment and Livestock Diseases) 58, 2–16.

Brown, Karen and Daniel Gilfoyle, eds. 2010. *Healing the Herds: Disease, Livestock Economies and the Globalization of Veterinary Medicine*. Athens: Ohio University Press.

Budiansky, Stephen. 1992. *The Covenant of the Wild: Why Animals Chose Domestication*. New York: W. Morrow.

Budiansky, Stephen. 2000. *The Truth About Dogs: An Inquiry into the Ancestry, Social Conventions, Mental Habits, and Moral Fiber of Canis Familiaris*. New York: Viking.

Bukatman, Scott. 1993. *Identity: The Virtual Subject in Postmodern Science Fiction*. Durham, N.C.: Duke University Press.

Burckhardt, Jacob. 1951. *The Civilization of the Renaissance in Italy*. Translated by S. G. C. Middlemore. London: Phaidon Press.

Burton, Maurice, and Robert Burton. 2002. *International Wildlife Encyclopedia*, Vol. 5. New York: Marshall Cavendish.

Cahill, James Leo. 2008. "Anacinema: Peter Tscherkassky's Cinematic Breakdowns: Towards the Unspeakable Film." *Spectator* 28, no. 2: 90–101.

Calarco, Matthew. 2008. *Zoographies: The Question of the Animal from Heidegger to Derrida*. New York: Columbia University Press.

Callon, Michel. 1986. "Some Elements of a Sociology of Transition: Domestication of the Scallops and the Fishermen of Saint Brieuc Bay." In *Power, Action and Belief: A New Sociology of Knowledge?* edited by John Law, 196–233. Boston: Routledge.

Campbell, Jeanne, and John Reesman. 1984. "Creatures of 'Charm': A New T. S. Eliot Poem." *Kenyon Review* 6:25–33.

Carmony, Neil B. 1995. *Onza! The Hunt for a Legendary Cat*. Silver City, N.M.: High Lonesome Books.

Carmony, Neil B., and David E. Brown, eds. 1991. *Mexican Game Trails: Americans Afield in Old Mexico, 1866–1940*. Norman: University of Oklahoma Press.

Carruthers, Jane. 1995. *The Kruger National Park: A Social and Political History*. Pietermaritzburg: University of Natal Press.

Carsten, Janet. 2003. *After Kinship*. Cambridge: Cambridge University Press.

Cartmill, Matt. 1993. *A View to a Death in the Morning: Hunting and Nature through History*. Cambridge, Mass.: Harvard University Press.

Cervantes, Miguel de. 1930. *Don Quixote*, Ozell's revision of the translation of Peter Motteux. New York: Modern Library.

Cervantes, Miguel de. 1998. "The Dialogue of Dogs." In *Exemplary Stories*, 250–305. Translated by Lesley Lipson. Oxford: Oxford University Press.

Chamberlin, J. Edward. 2002. "Hunting, Tracking and Reading." In *Literacy, Narrative and Culture*, edited by Jens Brockmeier, Min Wang, and David R. Olson, 67–85. Richmond, Surrey: Curzon Press.

Clark, Champ. 2008. "From Stray Dog to Movie Star: Rusco Leaves the Shelter and Finds a Home in the Hearts of Millions." *People*, November 3, 76.

Clinton, George. 1982. "Atomic Dog." First recorded in 1979 by Funk Record Co. by Capitol. http://www.duke.edu/~tmc/motherpage/list-singles.html.

CNN Talkback Live. 2002. Airing on March 6. Transcript #030600CN.V14, retrieved on October 3, 2003.

Coetzee, J. M. 2001. *The Lives of Animals*. Princeton, N.J.: Princeton University Press.

Cohen, Simona. 2009. *Animals as Disguised Symbols in Renaissance Art*. Boston: Brill.

Coleman, Jon T. 2004. *Vicious: Wolves and Men in America*. New Haven, Conn.: Yale University Press.

Conklin, Beth. 1993. "Hunting the Ancestors: Death and Alliance in Wari' Cannibalism." *Latin American Anthropology Review* 5, no. 2: 65–70.

Conklin, Beth. 2001. *Consuming Grief: Compassionate Cannibalism in an Amazonian Society*. Austin: University of Texas Press.

Cormier, Loretta. 2002. "Monkey as Food, Monkey as Child: Guajá Symbolic Cannibalism." In *Primates Face to Face: The Conservation Implications of Human-Nonhuman Primate Interconnections*, edited by Agustín Fuentes and Linda D. Wolfe, 63–84. Cambridge: Cambridge University Press.

Cosslett, Tess. 2002. "Child's Place in Nature: Talking Animals in Victorian Children's Fiction." *Nineteenth Century Contexts* 23:475–495.

Cosslett, Tess. 2006. *Talking Animals in British Children's Fiction, 1786–1914*. Aldershot, U.K.: Ashgate.

Cronon, William. 1999. "Speaking for Salmon." In *Making Salmon: An Environmental History of the Northwest Fisheries Crisis*, edited by Joseph E. Taylor, ix–xi. Seattle: University of Washington Press.

Cronon, William. 2003. *Changes in the Land: Indians, Colonists, and the Ecology of New England*. New York: Hill & Wang.

Crosby, Alfred. 1986. *Ecological Imperialism: The Biological Expansion of Europe, 900–1900*. Cambridge: Cambridge University Press.

Cummings, Brian. 2004. "Animal Language in Renaissance Thought." In *Renaissance Beasts: Of Animals, Humans, and Other Wonderful Creatures*, edited by Erica Fudge, 179–182. Urbana: University of Illinois Press.

Currier, Mary Jean P. 1983. "Felis concolor." *Mammalian Species* 200:1–7.

Curtis, Lewis Perry. 1971. *Apes and Angels: The Irishman in Victorian Caricature*. Devon: Newton Abbot, U.K: David & Charles.

Cuvier, Georges. 1817. *The Animal Kingdom Arranged in Conformity with Its Organization*, first English edition. London: Geo. B. Whittaker.

D'Amico, John F. 1983. *Renaissance Humanism in Papal Rome*. Baltimore: Johns Hopkins University Press.

Davies, Tony. 2008. *Humanism*. London: Routledge.

DeGette, Cara. 2002. "Public Eye: Free Obituaries; A Dying Breed." *Colorado Springs Independent*, April 10.

De Landa, Manuel. 1997. *A Thousand Years of Nonlinear History*. New York: Zone Books.

Derr, Mark. 2004. *A Dog's History of America: How Our Best Friend Explored, Conquered, and Settled a Continent*. New York: North Point Press.

Derrida, Jacques. 1992. *Given Time: I. Counterfeit Money*. Chicago: University of Chicago Press.

Derrida, Jacques. 1993. *Aporias: Dying—Awaiting (One Another At) the Limits of Truth*. Translated by Thomas Dutoit. Stanford, Calif.: Stanford University Press.

Derrida, Jacques. 1995. "'Eating Well,' or the Calculation of the Subject." In *Points . . . Interviews, 1974–1994*. Stanford, Calif.: Stanford University Press.

Derrida, Jacques. 1996. *Archive Fever: A Freudian Impression*. Chicago: University of Chicago Press.

Derrida, Jacques. 2008. *The Animal That Therefore I Am*. Translated by Marie-Louise Mallet. New York: Fordham University Press.

Derrida, Jacques, and Elisabeth Roudinesco. 2004. "Violence against Animals." In *For What Tomorrow . . . : A Dialogue.* Translated by Jeff Fort. Stanford, Calif.: Stanford University Press.

Desmond, Jane. 2003. "Relating to Animals after Death: Virtual Pet Cemeteries and the Public Act of Grieving Online." Paper presented at the International Anthrozoology Association conference at Kent State University, Canton, Ohio.

Dick, Philip K. 1996. "Man, Android and Machine." In *The Shifting Realities of Philip K. Dick: Selected Literary and Philosophical Writings*, edited by Lawrence Sutin, 211–232. New York: Vintage.

Dick, Philip K. 1999. *Do Androids Dream of Electric Sheep?* New York: Del Rey (also published in 1968).

DMX. 1999. "What's My Name?" on *And Then There Was X, Def Jam.* Lyrics by Irv Gotti. http://www.music.com/performance/whats_my_name/17.

Dolhinow, Phyllis. 2007. "Anthropology and Primatology." In *Primates Face to Face: The Conservation Implications of Human-Nonhuman Primate Interconnections*, edited by Agustín Fuentes and Linda D. Wolfe, 7–24. Cambridge: Cambridge University Press.

Doty, Mark. 2007. *Dog Years: A Memoir.* New York: HarperCollins.

Du Bellay, Joachim. 1931. *Poésies Françaises et Latines.* Edited by E. Courber. Paris: Librairie Garnier Frères.

Dubow, Jessica. 2000. "'From a View on the World to a Point of View in It': Rethinking Sight, Space and the Colonial Subject." *Interventions* 2, no. 1: 89–90.

Dubow, Jessica. 2004. "The Mobility of Thought: Reflections on Blanchot and Benjamin." *Interventions: The International Journal of Postcolonial Studies* 6, no. 2: 216–228.

Dubow, Jessica. 2004. "Out of Place and Other than Optical: Walter Benjamin and the Geography of Critical Thought." *Journal of Visual Culture* 3(3): 259–274.

Dubow, Jessica. 2008. "Minima Moralia: Or the Negative Dialectics of Exile." In *Imaginary Coordinates*, edited by R. Rosen, 56–63. Chicago: Spertus Institute of Jewish Studies.

Eisler, Colin. 1991. *Dürer's Animals.* Washington, D.C.: Smithsonian Institution Press.

Eliot, T. S. 1981. *Old Possum's Book of Practical Cats.* London: Faber and Faber (also published in 1939).

Emel, Jody. 1998. "Are You Man Enough, Big and Bad Enough? Wolf Eradication in the U. S." In *Animal Geographies: Place, Politics and Identity in the Nature-Culture Borderlands*, ed. J. Wolch and J. Emel. New York: Verso.

Erasmus, Desiderius. 1979. *The Praise of Folly.* Translated and edited by Clarence H. Miller. New Haven, Conn.: Yale University Press.

Eugenics Archive, Dolan DNA Learning Center, Cold Spring Harbor Laboratory. http://www.eugenicsarchive.org.

Farris, Gene. 2009. "PETA Dresses in KKK Garb Outside Westminster Dog Show." *USA Today*, February 10. http://www.usatoday.com/sports/2009–02–09-peta-westminster-kkk-protest_N.htm.

Fausto, Carlos. 2007. "Feasting on People: Eating Animals and Humans in Amazonia." *Current Anthropology* 48, no. 4: 497–530.

Fazion, Paolo. 1984. "L'Asino da leggere." In *Machiavelli: L'Asino e le bestie*, edited by Gian Mario Anselmi and Paolo Fazion, 25–134. Bologna: Cooperativa Libraria Universitaria Editrice.

Fleck, Ludwig. 1979. *The Genesis and Development of a Scientific Fact.* Chicago: University of Chicago Press.

Ford, John. 1971. *The Role of Trypanosomiases in African Ecology.* Oxford: Clarendon Press.

Foucault, Michel. 1991. *The Order of Things.* New York: Vintage.

Foucault, Michel. 1998. *The History of Sexuality: The Will to Knowledge*, Vol. 1. London: Penguin.

Freimer, Nelson, Ken Dewar, Jay Kaplan, and Lynn Fairbanks. n.d. "The Importance of the Vervet (African Green Monkey) as a Biomedical Model." http://www.genome.gov.

Fromm, Erich. 1997. *On Being Human*. New York: Continuum.

Frum, David. 2000. *How We Got Here: The '70s*. New York: Basic Books.

Fudge, Erica. 2000. *Perceiving Animals: Humans and Beasts in Early Modern English Culture*. London: Macmillan.

Fudge, Erica. 2002. *Animal*. London: Reaktion.

Fudge, Erica. 2002. "A Left-Handed Blow: Writing the History of Animals." In *Representing Animals*, edited by Nigel Rothfels, 3–18. Bloomington: Indiana University Press.

Fudge, Erica. 2006. *Brutal Reasoning: Animals, Rationality, and Humanity in Early Modern England*. Ithaca, N.Y.: Cornell University Press.

Fullwiley, Duana. 2008. "The Biologistical Construction of Race: 'Admixture' Technology and the New Genetic Medicine." *Social Studies of Science* 38, no. 5: 695–735.

Fulton, DoVeanna S. 2004. "Comic Views and Metaphysical Dilemmas: Shattering Cultural Images through Self-Definition and Representation by Black Comediennes." *Journal of American Folklore* 117:81–96.

Garin, Eugenio. 1965. *Italian Humanism: Philosophy and Civic Life in the Renaissance*. Westport, Conn.: Greenwood Press.

Gelli, Giambattista. 1549. *La Circe*. Florence: Torrentino.

Gelli, Giambattista. 1963. *The Circe of Signior Givanni Battista Gelli etc*. Translated by Thomas Brown with an introduction by Robert Adams. Ithaca, N.Y.: Cornell University Press.

Giblin, James. 1990. "Trypanosomiasis Control in African History: An Evaded Issue?" *Journal of African History* 31, no. 1: 59–80.

Gibson, William. 2000. *Neuromancer*. New York: Ace Trade Paperbacks (also published in 1984).

Gilbert, G. N., and M. Mulkay. 1984. *Opening Pandora's Box: A Sociological Analysis of Scientists' Discourse*. Cambridge: Cambridge University Press.

Gilfoyle, Daniel. 2006. "Veterinary Immunology as Colonial Science: Method and Quantification in the Investigation of Horsesickness in South Africa, c. 1905–1945." *Journal of the History of Medicine and Allied Sciences* 61, no. 1: 26–65.

Ginzburg, Carlo. 1989. *Clues, Myths, and the Historical Method*. Baltimore: Johns Hopkins University Press.

Gladwell, Malcolm. 2006. "Troublemakers: What Pit Bulls Can Teach Us about Profiling." *New Yorker*, February 6.

Glendinning, Simon. 1998. *On Being with Others: Heidegger, Derrida, Wittgenstein*. London: Routledge.

Glück, Louise. 1994. *Proofs & Theories: Essays on Poetry*. Hopewell, N.J.: Ecco Press.

Goldberg, Jonah. 1999. "Mau-Mauing the Dogcatcher: Is It Racist to Dislike a Dachshund?" *Slate.com*, March 11. http://www.slate.com/id/21547/.

Goldberg, Jonah. 2002. "Westminster Eugenics Show." *National Review Online*, February 13. http://article.nationalreview.com/?q=OTYyM2Y4YzEyNDJmYWIzNjNmYjE0M2NlY2MzYzlkMDA=.

Gontier, Thierry. 1998. *De l'homme à l'animal: Montaigne et Descartes ou les paradoxes de la philosophie moderne sur la nature des animaux*. Paris: Librairie philosophique J. Vrin.

Goodwin, John. 2004. "Jay-Z and Other Artists Need to Step Up against Dog Fighting." *U.S. Newswire* Op-Ed, October 12.

Gose, Elliott. 1988. *Mere Creatures: A Study of Modern Fantasy Tales for Children*. Toronto: University of Toronto Press.

Gould, Stephen Jay. "The Tooth and Claw Centennial." In *Dinosaur in a Haystack: Reflections in Natural History*, 63–75. New York: Three Rivers Press, 1996.

Grafton, Anthony, and Lisa Gardine. 1986. *From Humanism to the Humanities*. Cambridge, Mass.: Harvard University Press.

Grayson, Cecil, ed. and trans. 1983. "Il *Canis* di Leon Battista Alberti." In *Miscellanea di studi in onore di Vittore Branca, Umanesimo e Rinascimento a Firenze e Venezia*, Vol. 3 no. 1: 193–204. Florence: Leo Olschki.

Gregorian, Dareh. n.d. "Rapper DMX Mixed Up in Canine Suit." Foxnews.com. http://www.foxnews.com/story/0,2933,122436,00.html.

Grendler, Paul. 1989. *Schooling in Renaissance Italy: Literacy and Learning, 1300–1600*. Baltimore: Johns Hopkins University Press.

Grillo, Ernesto, ed. 1920. *Early Italian Literature*, Vol. 1, *Pre-Dante Poetical Schools*. London: Blackie.

Grossberger, Lewis. 2002. "Spot Dead! City Mourns," *MediaWeek* 12, no. 18 (June 6): 38.

Gudger, Eugene Willis. 1924. "Pliny's *Historia Naturalis*: the Most Popular Natural History Ever Published." *Isis* 6, no. 3: 269-281.

Hankins, James. 2003. "Two Twentieth-Century Interpreters of Renaissance Humanism: Eugenio Garin and Paul Oskar Kristeller." In *Humanism and Platonism in the Italian Renaissance*, Vol. 1, *Humanism*, 573–590. Rome: Edizioni di Storia e Letteratura.

Haraway, Donna J. 1989. *Primate Visions: Gender, Race, and Nature in the World of Modern Science*. New York: Routledge.

Haraway, Donna J. 1991. *Simians, Cyborgs, and Women: The Reinvention of Nature*. New York: Routledge.

Haraway, Donna J. 1997. *Modest_Witness@Second_Millennium.FemaleMan©Meets_OncoMouse™: Feminism and Technoscience*. New York: Routledge.

Haraway, Donna J. 2003. *The Companion Species Manifesto: Dogs, People, and Significant Otherness*. Chicago: Prickly Paradigm Press.

Haraway, Donna J. 2003. *The Haraway Reader*. New York: Routledge.

Haraway, Donna J. 2008. *When Species Meet*. Minneapolis: University of Minnesota Press.

Harbers, Hans. 2005. *Inside the Politics of Technology: Agency and Normativity in the Co-Production of Technology and Society*. Amsterdam: Amsterdam University Press.

Harbolt, Tami Lynne. 1993. "Too Loved to Be Forgotten: Pet Loss and Ritual Bereavement." Master's thesis, Western Kentucky University.

Harrison, Peter. 1998. "The Virtues of Animals in Seventeenth-Century Thought." *Journal of the History of Ideas* 59:463–484.

Hatten, James R., Anna Laura Averill-Murray, and William van Pelt. 2005. "A Spatial Model of Potential Jaguar Habitat in Arizona." *Journal of Wildlife Management* 69, no. 3: 1024–1033.

Hearne, Vicki. 1986. *Adam's Task: Calling Animals by Name*. New York: Knopf.

Hecht, Gabrielle. 1998. *The Radiance of France: Nuclear Power and National Identity after World War II*. Cambridge, Mass.: MIT Press.

Hecht, Gabrielle. 2002. "Rupture-talk in the Nuclear Age: Conjugating Colonial Power in Africa." *Social Studies of Science* 32, nos. 5–6: 691–728.

Hecht, Gabrielle. 2006. "Negotiating Global Nuclearities: Apartheid, Decolonization, and the Cold War in the Making of the IAEA." *Osiris* 21:25–48.

Heidegger, Martin. 1962. *Being and Time*. Translated by John Macquarrie and Edward Robinson. New York: Harper and Row.

Heidegger, Martin. 1977. *The Question Concerning Technology*. Translated by William Lovitt. New York: Garland (also published in 1954).

Heidegger, Martin. 1995. *The Fundamental Concepts of Metaphysics: World, Finitude, Solitude*. Bloomington: Indiana University Press.

Heise, Ursula K. 2003. "From Extinction to Electronics: Dead Frogs, Live Dinosaurs, and Electric

Sheep." In *Zoontologies: The Question of the Animal*, edited by Cary Wolfe, 59–81. Minneapolis: University of Minnesota Press.

Helmreich, Stefan. 2007. "An Anthropologist Underwater: Immersive Soundscapes, Submarine Cyborgs, and Transductive Ethnography." *American Ethnologist* 34, no. 4: 621–641.

Hemmings, Robert. 2007. "A Taste of Nostalgia: Children's Books from the Golden Age—Carroll, Grahame, and Milne." *Children's Literature* 35:54–79.

Hibben, Frank C. 1948. *Hunting American Lions*. Albuquerque: University of New Mexico Press.

Horowitz, Roger. 2003. "Making the Chicken of Tomorrow: Reworking Poultry as Commodities and as Creatures, 1945–1990." In *Industrializing Organisms: Introducing Evolutionary History*, edited by Philip Scranton and Susan R. Schrepfer, 215–235. New York: Routledge.

Howard, Cosmo. 2007. "Introducing Individualization." In *Contested Individualization: Debates About Contemporary Personhood*, edited by Cosmo Howard, 1–24. New York: Palgrave Macmillan.

Howard, Cosmo. 2007. "Three Models of Individualized Biography." In *Contested Individualization: Debates About Contemporary Personhood*, edited by Cosmo Howard, 25–44. New York: Palgrave Macmillan.

Humane Society of the United States (HSUS). 2008. "Online Card Thanks Future First Family for Choosing to Adopt a Dog." Press release, November 25. http://www.hsus.org/press_and _publications/press_releases/online_card_thanks_obamas_for_choosing_adoption_112508.html.

Humane Society of the United States (HSUS). 2008. "Second Chance Dog Heads to the White House." Press release, April 12. http://www.hsus.org/pets/pets_related_news_and_events/second -chance_dog_heads_to.html.

Hume, Janice. 2000. *Obituaries in American Culture*. Jackson: University Press of Mississippi.

Hunt, Nancy R. 1999. *A Colonial Lexicon: Of Birth Ritual, Medicalization, and Mobility in the Congo. (Body, Commodity, Text)*. Durham, N.C.: Duke University Press.

Ingold, Tim. 1974. "On Reindeer and Men." *Man* 9, no. 4: 523–538.

Ingold, Tim. 2000. *The Perception of the Environment: Essays in Livelihood, Dwelling, and Skill*. New York: Routledge.

Irvine, Leslie. 2004. *If You Tame Me: Understanding Our Connection with Animals*. Philadelphia: Temple University Press.

Jasanoff, Sheila. 2006. *States of Knowledge: The Co-Production of Science and the Social Order*. London: Routledge.

Jenkins, Mark. 2008. "Who Murdered the Virunga Gorillas?" *National Geographic* 214, no. 1: 34–65.

Johnson, Amrah Salomon. 2005. "Is the Bandit Back Disguised as a Chihuahua?" *El Tecolote Online*, March 7. http://news.eltecolote.org/news/view_article.html?article_id=f337b681beb06d75abffc2 d06a03e7f4.

Johnson, Marilyn. 2007. *The Dead Beat: Lost Souls, Lucky Stiffs, and the Perverse Pleasures of Obituaries*. New York: Harper Perennial.

Johnson, Terry B., William E. Van Pelt, and James N. Stuart. 2009. *Jaguar Conservation Assessment for Arizona, New Mexico, and Northern Mexico*. Phoenix: Arizona Game and Fish Department.

Kalof, Linda. 2007. *Looking at Animals in Human History*. London: Reaktion.

Kalof, Linda, and Amy Fitzgerald, eds. 2007. *The Animals Reader: The Essential Classic and Contemporary Writings*. Oxford: Berg.

Katz, Jon. 2003. *The New Work of Dogs: Tending to Life, Love and Family*. New York: Villard.

Keith, Christie. 2009. "It's Just Puppy Love: Why All Dog Lovers Should Be Happy About the Obamas' New Dog." *San Francisco Gate*, April 14. http://www.sfgate.com/cgi-bin/article.cgi?f=/ g/a/2009/04/14/petsc01041409.DTL.

Kennedy, Patricia F., and Mary G. McGarvey. 2008. "Animal-Companion Depictions in Women's Magazine Advertising." *Journal of Business Research* 61:424–430.

Kerr, Owen. 1976. "*Old Possum's Book of Practical Cats* by T. S. Eliot." *English Journal* 65, no. 1: 66.

Kilgour, Maggie. 1998. "The Function of Cannibalism at the Present Time." In *Cannibalism and the Colonial World*, edited by Francis Barker, Peter Hulme, and Margaret Iversen, 238–259. Cambridge: Cambridge University Press.

Kittler, Friedrich A. 1999. *Gramophone, Film, Typewriter*. Stanford, Calif.: Stanford University Press.

Kjekshus, Helge. 1996. *Ecology Control and Economic Development in East African History*, 2nd ed. London: James Currey.

Klingender, Francis. 1971. *Animals in Art and Thought to the End of the Middle Ages*, edited by Evelyn Antal and John Harthan. Cambridge, Mass.: MIT Press.

Knobloch, Frieda. 2001. Review of *Making Salmon: An Environmental History of the Northwest Fisheries Crisis* by Joseph E. Taylor III. *American Historical Review* 106, no. 4: 1373–1374.

Köhler, Axel. 2005. "Of Apes and Men: Baka and Bantu Attitudes to Wildlife and the Making of Eco-Goodies and Baddies." *Conservation and Society* 3, no. 2: 407–435.

Kohn, Eduardo. 2007. "How Dogs Dream: Amazonian Natures and the Politics of Transpecies Engagement." *American Ethnologist* 34, no. 1: 3–24.

Konigsberg, Eric. 2008. "Beloved Pets Everlasting?" *New York Times*, December 31. http://www.nytimes.com/2009/01/01/garden/01clones.html.

Kriger, Norma J. 1992. *Zimbabwe's Guerrilla War: Peasant Voices*. Cambridge: Cambridge University Press.

Kristeller, Paul Oskar. 1979. *Renaissance Thought and Its Sources*. New York: Columbia University Press.

Kristeller, Paul Oskar. 1988. "Humanism." In *The Cambridge History of Renaissance Philosophy*, edited by Charles B. Schmitt and Quentin Skinner, 113–137. Cambridge: Cambridge University Press.

Kuhn, Thomas. 1962. *The Structure of Scientific Revolutions*. Chicago: University of Chicago Press.

Kurtén, Bjorn, and Elaine Anderson. 1980. *Pleistocene Mammals*. New York: Columbia University Press.

Langwick, Stacey. 2007. "Devils, Parasites, and Fierce Needles: Healing and the Politics of Translation in Southern Tanzania." *Science, Technology & Human Values* 32, no. 1: 88–117.

Langwick, Stacey. 2008. "Articulate(d) Bodies: Traditional Medicine in a Tanzanian Hospital." *American Ethnologist* 35, no. 3: 428–439.

Latour, Bruno. 1987. *Science in Action: How to Follow Scientists and Engineers through Society*. Cambridge, Mass.: Harvard University Press.

Latour, Bruno. 1988. *The Pasteurization of France*. Cambridge, Mass.: Harvard University Press.

Latour, Bruno. 1993. *We Have Never Been Modern*. Translated by Catherine Porter. Cambridge, Mass.: Harvard University Press.

Latour, Bruno. 1996. *Aramis, or the Love of Technology*. Cambridge, Mass.: Harvard University Press.

Latour, Bruno. 1999. *Pandora's Hope: Chapters on the Reality of Science Studies*. Cambridge, Mass.: Harvard University Press.

Latour, Bruno. 2004. *Politics of Nature: How to Bring the Sciences into Democracy*. Translated by Catherine Porter. Cambridge, Mass.: Harvard University Press.

Latour, Bruno. 2005. *Reassembling the Social: An Introduction to Actor-Network Theory*. New York: Oxford University Press.

Latour, Bruno, and Peter Weibel, eds. 2005. *Making Things Public: Atmospheres of Democracy*. Cambridge, Mass.: MIT Press.

Latour, Bruno, and Steve Woolgar. 1986. *Laboratory Life: The Construction of Scientific Facts*, 2nd ed. Princeton, N.J.: Princeton University Press (also published in 1979).

Lee, Paula Young, ed. 2008. *Meat, Modernity, and the Rise of the Slaughterhouse*. Durham: University of New Hampshire Press.

Leopold, A. Starker. 1959. *Wildlife in Mexico: The Game Birds and Mammals*. Berkeley: University of California Press.

Leopold, Aldo. 1949. *A Sand County Almanac*. New York: Oxford University Press.

Levey, Bob. 2002. "Latest in Questionable Taste: Pet Obituaries." *Washington Post*, May 1.

Levinas, Emmanuel. 1998. *Of God Who Comes to Mind*. Translated by Bettina Bergo. Stanford, Calif.: Stanford University Press.

Liebenberg, Louis. 1990. *The Art of Tracking: The Origin of Science*. Claremont, South Africa: David Philip.

Lippit, Akira Mizuta. 2002. ". . . From Wild Technology to Electric Animal." In *Representing Animals*, edited by Nigel Rothfels, 119–136. Bloomington: Indiana University Press.

Long, George. 1839. *The Penny Cyclopedia of the Society of Useful Knowledge*, Vol. 13. London: Charles Knight.

Love, Glen. 1996. "Revaluing Nature: Toward an Ecological Criticism." In *The Ecocriticism Reader: Landmarks in Literary Ecology*, edited by Cheryll Glotfelty and Harold Fromm, 225–240. Athens: University of Georgia Press.

Lovell-Smith, Rose. 2007. "Eggs and Serpents: Natural History Reference in Lewis Carroll's Scene of Alice and the Pigeon." *Children's Literature* 35:27–53.

Loviglio, Joann. 2002. "Philadelphia Daily News Adding Pet Death Notices." Associated Press, March 5.

Lowood, Henry. 1995. "The New World and the European Catalog of Nature." In *America in European Consciousness 1493–1750*, edited by Karen Ordahl Kupperman, 295–323. Chapel Hill: University of North Carolina Press.

Luckert, Karl. 1975. *The Navajo Hunter Tradition*. Tucson: University of Arizona Press.

Lundin, Anne H. 1994. "Victorian Horizons: The Reception of Children's Books in England and America, 1880–1900." *Library Quarterly* 64, no. 1 (January): 30–59.

Lutts, Ralph H., ed. 1998. *The Wild Animal Story*. Philadelphia: Temple University Press.

Lutts, Ralph H. 2001. *The Nature Fakers: Wildlife, Science and Sentiment*. Charlottesville: University Press of Virginia.

Lyons, Maryinez. 1988. "Sleeping Sickness Epidemics and Public Health in the Belgian Congo." In *Imperial Medicine and Indigenous Societies*, edited by David Arnold, 105–124. Manchester: Manchester University Press.

Lyons, Maryinez. 1991. "African Sleeping Sickness: An Historical Review." *International Journal of STD & AIDS* Supplement 1, no. 2: 20–25.

Lyons, Maryinez. 1992. *The Colonial Disease: A Social History of Sleeping Sickness in Northern Zaire, 1900–1940*. Cambridge: Cambridge University Press.

Machiavelli, Niccolò. 1960. "Dell'Asino d'oro." In *Tutte le opere*, Vol. 2, edited by Francesco Flora and Carlo Cordiè, 751–781. Milano: A. Mondadori.

Machiavelli, Niccolò. 1961. *Lettere*. Edited by Franco Gaetà. Milan: Feltrinelli.

Mackenzie, Adrian. 2002. *Transductions: Bodies and Machines at Speed*. New York: Continuum.

Mackenzie, Donald, and Judy Wajcman, eds. 1985. *The Social Shaping of Technology*. Buckingham, U.K.: Open University Press.

MacKenzie, John M. 1990. "Experts and Amateurs: Tsetse, Nagana and Sleeping Sickness in East and Central Africa." In *Imperialism and the Natural World*, edited by John M. MacKenzie, 187–212. Manchester: Manchester University Press.

Malamud, Randy. 2003. *Poetic Animals and Animal Souls*. New York: Palgrave.

Malinowski, Bronislaw. 1961. *Argonauts of the Western Pacific: An Account of Native Enterprise and Adventure in the Archipelagoes of Melanesian New Guinea*. New York: E. P. Dutton.

Malowany, Maureen. 2000. "Unfinished Agendas: Writing the History of Medicine of Sub-Saharan Africa." *African Affairs* 99:325–349.

Mangum, Teresa. 2007. "Animal Angst: Victorians Memorialize Their Pets." In *Victorian Animal Dreams: Representations of Animals in Victorian Literature*, edited by Deborah Denenholz Morse and Martin A. Danahay, 15–34. Hampshire, U.K.: Ashgate.

Marius, Richard. 1999. *Thomas More: A Biography*. Cambridge, Mass.: Harvard University Press.

Marks, Alan, and Tommy Piggee. 1998/1999. "Obituary Analysis and Describing a Life Lived: The Impact of Race, Gender, Age, and Economic Status." *Omega* 38, no. 1: 37–57.

Mason, Jim. 1993. *An Unnatural Order: Why We Are Destroying the Planet and Each Other*. New York: Continuum.

Masson, Jeffrey Moussaieff, and Susan McCarthy. 2007. "Grief, Sadness, and the Bones of Elephants." In *The Animals Reader: The Essential Classic and Contemporary Writings*, edited by Linda Kalof and Amy Fitzgerald, 91–103. Oxford: Berg.

Matthiessen, Peter. 1959. *Wildlife in America*. New York: Viking Press.

Maulde la Clavière, René de. 1900. *The Women of the Renaissance*. London: J. Allen.

Mauss, Marcel. 1990. *The Gift: The Form and Reason for Exchange in Archaic Societies*. New York: W. W. Norton (also published in 1950).

Mavhunga, Clapperton. 2006. "Big Game Hunters, Bacteriologists, and Tsetse Fly Entomology in Colonial South-East Africa: The Selous-Austen Debate Revisited, 1905–1940s." *ICON* 12:75–117.

Mavhunga, Clapperton. 2008. "The Mobile Workshop: Mobility, Technology, and Human-Animal Interaction in Gonarezhou (National Park), 1850–Present." Ph.D. diss., University of Michigan.

Mavhunga, Clapperton and Marja Spierenburg. 2009. "Transfrontier Talk, Cordon Politics: The Early History of the Great Limpopo Transfrontier Park in Southern Africa, 1925–1940." *Journal of Southern African Studies* 35(3): 715–735.

Maybury, Karol K. 1995/1996. "Invisible Lives: Women, Men, and Obituaries." *Omega* 32, no. 1: 27–37.

McCain, Emil B., and Jack L. Childs. 2008. "Evidence of Resident Jaguars (*Panthera onca*) in the Southwestern United States and the Implications for Conservation." *Journal of Mammalogy* 89, no. 1: 1–10.

McCann, James C. 1999. *Green Land, Brown Land, Black Land: An Environmental History of Africa*. Portsmouth, N.H.: Heinemann.

McCann, James C. 2005. *Maize and Grace: Africa's Encounter with a New World Crop, 1500–2000*. Cambridge, Mass.: Harvard University Press.

McHugh, Susan. 2002. "Bitches from Brazil: Cloning and Owning Dogs through the Missyplicity Project." In *Representing Animals*, edited by Nigel Rothfels, 180–198. Bloomington: Indiana University Press.

McKelvey, John J. 1973. *Man against Tsetse: Struggle for Africa*. Ithaca, N.Y.: Cornell University Press.

McKinney, Debra. 2002. "Newspaper Obituaries Pay Tribute to Pets." *Anchorage Daily News*, April 7.

McNeil, Donald G., Jr. 1999. "The Great Ape Massacre." *New York Times*, May 9.

McNeill, J. R. 1985. *The Atlantic Empires of France and Spain: Louisbourg and Havana, 1700–1763*. Chapel Hill: University of North Carolina Press.

McNeill, J. R. 1992. *The Mountains of the Mediterranean World: An Environmental History*. New York: Cambridge University Press.

McNeill, J. R. 2003. "Observations on the Nature and Culture of Environmental History." *History and Theory* (December): 5–43.

Melson, Gail. 2001. *Why The Wild Things Are*. Cambridge, Mass.: Harvard University Press.

Menache, Sophia. 2000. "Hunting and Attachment to Dogs in the Pre-Modern Period." In *Companion Animals and Us: Exploring the Relationships between People and Pets*, edited by Anthony L. Podberscek, Elizabeth S. Paul, and James A. Serpell, 42–60. Cambridge: Cambridge University Press.

Mendelsohn, J. Andrew. 1998. "From Eradication to Equilibrium: How Epidemics Became Complex after World War I." In *Greater Than the Parts: Holism in Bio-medicine 1920–1950*, edited by Christopher Lawrence and George Weisz, 303–331. Oxford: Oxford University Press.

Merleau-Ponty, Maurice. 1945. *Phenomenology of Perception*. New York: Routledge.

Midgley, Mary. 1983. *Animals and Why They Matter*. Athens: University of Georgia Press.

Mighetto, Lisa. 1991. *Wild Animals and American Environmental Ethics*. Tucson: University of Arizona Press.

Millais, Sir Everett. 1895. *Two Problems of Reproduction: Lecture Delivered at St. Thomas' Hospital on February 25, 1895*. Manchester/London: Our Dogs' Publishing.

Mitani, John C., William J. Sanders, Jeremiah S. Lwanga, and Tammy L. Windfelder. 2001. "Predatory Behavior of Crowned Hawk-eagles (*Stephanoaetus coronatus*) in Kibale National Park, Uganda." *Behavioral and Ecological Sociobiology* 49:187–195.

Mitchell, Timothy. 2003. *Rule of Experts: Egypt, Techno-Politics, Modernity*. Berkeley: University of California Press.

Mithen, Steven J. 1990. *Thoughtful Foragers: A Study of Prehistoric Decision Making*. New York: Cambridge University Press.

Molella, Arthur, and Joyce Bedi, eds. 2003. *Inventing for the Environment*. Cambridge, Mass.: MIT Press.

Montaigne, Michel de. 1991. *The Complete Essays*. Translated by M. A. Screech. London: Penguin.

More, Thomas. 2002. *Utopia*, rev. ed. Translated by Robert M. Adams and edited by George M. Logan and Robert M. Adams. Cambridge: Cambridge University Press.

Moremem, Robin D., and Cathy Cradduck. 1998/1999. "'How Will You Be Remembered After You Die?' Gender Discrimination after Death Twenty Years Later." *Omega* 38, no. 4: 241–254.

Moseman, Andrew. 2008. "Thanks to His Own Popularity, Nemo Can't Be Found." Blog, Discovermagazine.com, June 27. http://blogs.discovermagazine.com/discoblog/2008/06/27/thanks-to-his-own-popularity-nemo-cant-be-found/.

Mosteller, Jill. 2008. "Animal-Companion Extremes and Underlying Consumer Themes." *Journal of Business Research* 61:512–521.

Murray, Martin. 1995. "Blackbirding at Crooks' Corner: Illicit Labor Recruiting in the Northeastern Transvaal, 1910–1940." *Journal of Southern African Studies* 21, no. 3: 373–397.

Nadasdy, Paul. 2007. "The Gift in the Animal: The Ontology of Hunting and Human-Animal Sociality." *American Ethnologist* 34, no. 1: 25–43.

Nash, Linda. 2005. "The Agency of Nature or the Nature of Agency?" *Environmental History* 10, no. 1: 67–69.

Newmyer, Stephen T. 2007. "Animals in Ancient Philosophy: Conceptions and Misconceptions." In *A Cultural History of Animals in Antiquity*, 151–174, edited by Linda Kalof. Oxford: Berg.

Niehaus, Isak. 2002. "Ethnicity and the Boundaries of Belonging: Reconfiguring Shangaan Identity in the South African Lowveld." *African Affairs* 101:557–583.

Nietzsche, Friedrich. 1983. "On the Uses and Disadvantages of History for Life." In *Untimely Meditations*, 57–123. Translated by R. J. Hollingdale and edited by Daniel Breazeale. Cambridge: Cambridge University Press.

Nietzsche, Friedrich. 1997. *Untimely Meditations*. Translated by R. J. Hollingdale and edited by Daniel Breazeale. Cambridge: Cambridge University Press.

Nowak, Ronald M. 1999. *Walker's Mammals of the World*, 6th ed. Baltimore: Johns Hopkins University Press.

O'Neill, Charles A. T., ed. 1985. *The American Kennel Club: 1884–1984*. New York: Howell Book House.

Orr, Jimmy. 2009. "Obama's Dog Bo: Cute but What About the Promise to Adopt?" *Christian Science Monitor Online*, April 13. http://features.csmonitor.com/politics/2009/04/13/obamas-dog-bo-cute-but-what-about-promise-to-adopt/.

Papy, Jan. 1999. "Lipsius and His Dogs: Humanist Traditions, Iconography and Rubens's Four Philosophers." *Journal of the Warburg and Courtault Institutes* 62:167–198.

Pavlik, Steve. 2003. "Rohonas and Spotted Lions: The Historical and Cultural Occurrence of the Jaguar, *Panthera onca*, among the Native Tribes of the American Southwest." *Wicazo Sa Review* 18, no. 1: 157–175.

People for the Ethical Treatment of Animals (PETA). n.d. "Why Is the AKC Like the KKK?" http://cdm.amphilsoc.org/cdm4/results.php?CISOOP1=exact&CISOFIELD1=CISOSEARCHALL&CISOROOT=/eugenics&CISOBOX1=Mss.+575.06+Am3.

Perfetti, Stefano. 2007. "Philosophers and Animals in the Renaissance." In *A Cultural History of Animals in the Renaissance*, edited by Bruce Boehrer, 147–164. Oxford: Berg.

Perissinotto, Luigi. 1998. "Perché gli animali spesso usano la ragione meglio dell'uomo." *Rivista di estetica*, n.s. 8, no. 38: 177–196.

Perkins, Agnes. 1980. "Critical Summary of Recent Journal Articles on Poetry for Children." *Children's Literature Association Quarterly* 5, no. 2: 35–38.

Perry, Richard. 1970. *The World of the Jaguar*. New York: Taplinger.

Peterson, Dale. 2003. *Eating Apes*. Berkeley: University of California Press.

Peterson, Tom. 2001. "Media: Your Fame, Their Fortune; C-J Moves to Paid Obituaries." *LEO: Louisville Eccentric Observer*, July 25, 13.

Petrarch. 1966. *Selected Sonnets, Odes and Letters*. Edited by Thomas G. Bergin. New York: Appleton-Century-Crofts.

Phebus, Gaston. 1998. *The Hunting Book of Gaston Phebus: Manuscrit Français 616*. Edited by Wilhelm Schlag with an introduction by Marcel Thomas and Francois Avril. Paris: Bibliothèque Nationale.

Phoofolo, Pule. 1993. "Epidemics and Revolutions: The Rinderpest Epidemic in Late Nineteenth-Century Southern Africa." *Past & Present* 138, no. 1: 112–143.

Phoofolo, Pule. 2003. "Face to Face with Famine: The BaSotho and the Rinderpest, 1897–1899." *Journal of Southern African Studies* 29, no. 2: 503–527.

Pickering, A. 1984. *Constructing Quarks: A Sociological History of Particle Physics*. Chicago: University of Chicago Press.

Plutarch. 1995. "Beasts Are Rational." In *Moralia*, Vol. 12, edited by G. P. Goold, 489–533. Cambridge, Mass.: Harvard University Press.

Plutarch. 1995. "On the Eating of Flesh, I-II." In *Moralia*, Vol. 12, edited by G. P. Goold, 537–579. Cambridge, Mass.: Harvard University Press.

Plutarch. 1995. "Whether Land or Sea Animals Are Cleverer." In *Moralia*, Vol. 12, edited by G. P. Goold, 311–479. Cambridge, Mass.: Harvard University Press.

Pollan, Michael. 2001. *Botany of Desire: A Plant's Eye View of the World*. New York: Random House.

Pompilio, Natalie. 2002. "The Brides Wore White." *American Journalism Review* (October): 12.

Porphyry of Tyre. 2000. *On Abstinence from Killing Animals*. Translated by Gillian Clark. London: Duckworth.

Rabil, Albert, Jr. 1981. *Laura Cereta, Quattrocento Humanist*. Binghamton, N.Y.: Center for Medieval and Early Renaissance Studies.

Rabinowitz, Alan R. 1999. "The Present Status of Jaguars (*Panthera onca*) in the Southwestern United States." *Southwestern Naturalist* 44, no. 1: 96–100.

Rabinowitz, Alan R., and B. G. Nottingham Jr. 1986. "Ecology and Behavior of the Jaguar (*Panthera onca*) in Belize, Central America." *Journal of Zoology* 210, no. 1: 149–159.

Raby, Frederic J. E. 1967. *A History of Secular Latin Poetry in the Middle Ages*, 2nd. ed., Vol. 2. Oxford: Clarendon Press.

Ramutsindela, Maano F. 2002. "The Perfect Way to Ending a Painful Past? Makuleke Land Deal in South Africa." *Geoforum* 32:15–24.

Ranger, Terence Osborne. 1999. *Voices from the Rocks: Nature, Culture & History in the Matopos Hills of Zimbabwe*. Bloomington: Indiana University Press.

Resl, Brigitte. 2007. "Introduction: Animals in Culture, ca. 1000–ca. 1400." In *A Cultural History of Animals in the Middle Ages*, edited by Brigitte Resl, 1–26. Oxford: Berg.

Reyes, Ernesto Alvarado. 2008. "The Legend of the Mexican Onza." *Mastozoologia Neotropical* 15, no. 1: 147–148.

Rilke, Rainer Maria. 1978. *Duino Elegies*. New York: Norton.

Ritvo, Harriet. 1987. *The Animal Estate: The English and Other Creatures in the Victorian Age*. Cambridge, Mass.: Harvard University Press.

Ritvo, Harriet. 2004. "Animal Planet." *Environmental History* 9, no. 2: 204–220.

Roberts, Michael. 2002. "Dead Lines: The Denver Dailies Change the Way They Handle Obituaries—For Better and for Worse." *Westword*, April 18. http://rss.westword.com/2002–04–18/news/dead-lines/1.

Roosevelt, Theodore. 1893. "Hunting the Grisly and Other Sketches." In *The Works of Theodore Roosevelt*, vol. 13. New York: P. F. Collier and Son.

Rorarius, Hieronymus. 1648. *Quod animalia bruta saepe ratione utantur melius homine*. Edited by Gabriel Naudé. Paris: Cramoisy.

Rose, Dale, and Stuart Blume. 2003. "Citizens as Users of Technology: An Exploratory Study of Vaccines and Vaccination." In *How Users Matter*, edited by Nelly Oudshoorn and Trevor Pinch, 105–131. Cambridge, Mass.: MIT Press.

Rothfels, Nigel, ed. 2002. *Representing Animals*. Bloomington: Indiana University Press.

Rubin, Diana, ed. and trans. 1997. *The Collected Letters of a Renaissance Feminist*. Chicago: University of Chicago Press.

Russell, Edmund P. 2001. *War and Nature: Fighting Humans and Insects with Chemicals from World War I to Silent Spring*. Cambridge: Cambridge University Press.

Russell, Edmund. 2003. "Evolutionary History: Prospectus for a New Field." *Environmental History* 8, no. 2: 204–228.

Ryder, Richard D. 1989. *Animal Revolution: Changing Attitudes towards Speciesism*. Oxford: Basil Blackwell.

Rymes, Betsy. 1996. "Naming as Social Practice: The Case of Little Creeper from Diamond Street." *Language in Society* 25, no. 2: 237–260.

Safina, Carl. 2002. *Eye of the Albatross: Visions of Hope and Survival*, Vol. 1. New York: Henry Holt.

Salisbury, Joyce E. 1994. *The Beast Within: Animals in the Middle Ages*. New York: Routledge.

Sambucus, Johannes [János Zsámboky]. 1564. *Emblemata*. Antwerp: C. Plantin.

Sanders, Clinton. 1999. *Understanding Dogs: Living and Working with Canine Companions*. Philadelphia: Temple University Press.

Sanders, Clinton R. and Arluke, Arnold. 2007. "Speaking for Dogs." In *The Animals Reader: The Essential Classic and Contemporary Writings*, edited by Linda Kalof and Amy Fitzgerald, 63–69. Oxford: Berg.

Sanderson, Eric W., Kent H. Redford, Cheryl-Lesley B. Chetkiewicz, Rodrigo A. Medellin, Alan

R. Rabinowitz, John G. Robinson, and Andrew B. Taber. 2002. "Planning to Save a Species: The Jaguar as a Model." *Conservation Biology* 16, no. 1: 58–72.

Saunders, Nicholas J. 1994. "Predators of Culture: Jaguar Symbolism and Mesoamerican Elites." *World Archaeology* 26, no. 1: 104–117.

Saunders, Nicholas J. 1998. "Architecture of Symbolism: The Feline Image." In *Icons of Power: Feline Symbolism in the Americas*, edited by Nicholas Saunders, 14–15. New York: Routledge.

Saunders, Nicholas J., ed. 1998. *Icons of Power: Feline Symbolism in the Americas*. New York: Routledge.

Sax, Boria. 2007. "Animals as Tradition." In *The Animals Reader: The Essential Classic and Contemporary Writings*, edited by Linda Kalof and Amy Fitzgerald, 270–277. Oxford: Berg.

Scala, Aidée. 2004. *Girolamo Rorario: Un umanista diplomatico e i suoi dialoghi*. Florence: Olschki.

Schaller, George B., and Peter Gransden Crawshaw Jr. 1980. "Movement Patterns of Jaguar." *Biotropica* 12, no. 3: 161–168.

Schmitt, Jean-Claude. 1983. *The Holy Greyhound: Guinefort, Healer of Children since the Thirteenth Century*. Translated by Martin Thom. Cambridge: Cambridge University Press.

Schneider, David. 1980. *American Kinship: A Cultural Account*. Chicago: University of Chicago Press.

Scranton, Philip, and Susan R. Schrepfer, eds. 2003. *Industrializing Organisms: Introducing Evolutionary History*. New York: Routledge.

Semelin, Jacques. 2007. *Purify and Destroy: The Political Uses of Massacre and Genocide*. New York: Columbia University Press.

Sennett, Richard. 2006. *The Culture of the New Capitalism*. New Haven, Conn.: Yale University Press.

Serpell, James. 1986. *In the Company of Animals: A Study of Human-Animal Relationships*. Oxford: Basil Blackwell.

Seton, Ernest Thompson. 1909. "The Oldest of All Writing—Tracks." *Country Life in America* 17 (December): 169–173, 242, 244, 246, 248, 250.

Seymour, Kevin L. 1989. "*Panthera onca*." *Mammalian Species* 340:1–9.

Sheeler, Jim. 2007. *Obit: Inspiring Stories of Ordinary People Who Led Extraordinary Lives*. New York: Penguin Books.

Shell, Hanna Rose. 2004. "Skin Deep: American Taxidermy, Embodiment and Extinction." In *The Past, Present & Future of Natural History Museums*, edited by A. Leviton, 88–112. San Francisco: Academy of Sciences.

Shell, Hanna Rose. 2005. "Things Under Water: Etienne-Jules Marey's Aquarium Laboratory and Cinema's Assembly." In *Dingpolitik: Atmospheres of Democracy*, edited by Bruno Latour and Peter Weibel, 326–332. Cambridge, Mass.: MIT Press.

Shelton, Jo-Ann. 2007. "Beastly Spectacles in the Ancient Mediterranean World." In *A Cultural History of Animals in Antiquity*, edited by Linda Kalof, 97–126. Oxford: Berg.

Shepard, Paul. 1973. *The Tender Carnivore and the Sacred Game*. New York: Scribner.

Shepard, Paul. 1996. *The Others: How Animals Made Us Human*. Washington, D.C.: Island Press.

Simonian, Lane. 1995. *Defending the Land of the Jaguar: A History of Conservation in Mexico*. Austin: University of Texas Press.

Simons, John. 2002. *Animal Rights and the Politics of Literary Representation*. New York: Palgrave.

Singer, Peter. 1990. *Animal Liberation*, 2nd ed. New York: New York Review.

Singer, Peter. 2001. "A Utilitarian Defense of Animal Liberation." In *Environmental Ethics*, edited by Louis Pojman, 33–39. Stamford, Conn.: Wadsworth.

Singer, Peter, and Lori Gruen. 1987. *Animal Liberation: A Graphic Guide*. London: Camden Press.

Skabelund, Aaron. 2005. "Can the Subaltern Bark? Imperialism, Civilization, and Canine Cultures in Nineteenth-Century Japan." In *JAPANimals: History and Culture in Japan's Animal Life*, edited by

Gregory M. Pflugfelder and Brett L. Walker, 195–243. Ann Arbor: Center for Japanese Studies, University of Michigan.

Slater, Candace. 1994. *Dance of the Dolphin: Transformation and Disenchantment in the Amazonian Imagination*. Chicago: University of Chicago Press.

Smith, Frederick. 1976. *The Early History of Veterinary Literature and Its British Development*, Vol. 1, *From the Earliest Period to a.d. 1700*. London: J. A. Allen.

Sobol, Peter G. 1993. "The Shadow of Reason: Explanations of Intelligent Animal Behavior in the Thirteenth Century." In *The Medieval World of Nature*: *A Book of Essays*, edited by Joyce E. Salisbury, 109–128. New York: Garland.

Sorabji, Richard. 1993. *Animal Minds and Human Morals: The Origin of the Western Debate*. Ithaca, N.Y.: Cornell University Press.

Spears, Nancy E., John C. Mowen, and Goutam Chakraborty. 1996. "Symbolic Role of Animals in Print Advertising: Content Analysis and Conceptual Development." *Journal of Business Research* 37, no. 2: 87–95.

Spiegel, Marjorie. 1996. *The Dreaded Comparison: Human and Animal Slavery*. New York: Mirror Books.

Spila, Cristiano, ed. 2002. *Cani di pietra: L'epicedio canino nella poesia del Rinascimento*. Translated by Maria Gabriella Critelli and Cristiano Spila. Rome: Quiritta.

Squier, Susan. 2009. "Fellow-Feeling." In *Animal Encounters*, edited by Tom Tyler and Manuela Rossini, 173–196. Boston: Brill.

Stables, Gordon. 1877. *The Practical Kennel Guide, with Plain Instructions How to Rear and Breed Dogs for Pleasure, Show, and Profit*. London: Cassell Petter & Galpin.

Staden, Hans. 2008. *Hans Staden's True History: An Account of Cannibal Captivity in Brazil*. Durham, N.C.: Duke University Press.

Stark, Nigel. 2006. *Life After Death: The Art of the Obituary*. Victoria, Australia: Melbourne University Publishing.

Steenkamp, Conrad, and J. Uhr. 2000. *The Makuleke Land Claim: Power Relations and Community-Based Natural Resource Management*. London: International Institute for Environment and Development.

Stengers, Isabelle. 2002. *Penser avec Whitehead: Une Libre et Sauvage Création de Concepts*. Paris: Seuil.

Storey, William Kelleher. 2008. *Guns, Race, and Power in Colonial South Africa*. Cambridge: Cambridge University Press.

Strathern, Marilyn. 1988. *The Gender of the Gift: Problems with Women and Problems with Society in Melanesia*. Berkeley: University of California Press.

Strathern, Marilyn. 1992. *After Nature: English Kinship in the Late Twentieth Century*. Cambridge: Cambridge University Press.

Sullivan, Carl. 2002. "Same-Sex Unions Gain Notice; But Papers Not Wedded to Policies." *Editor and Publisher Magazine*, September 2, 6.

Suutala, Maria. 1990. *Tier und Mensch im Denken der Deutschen Renaissance*. Helsinki: SHS.

Swynnerton, C. F. M. 1921. "An Examination of the Tsetse Problem in North Mossurise, Portuguese East Africa." *Bulletin of Entomological Research* 11:304–330.

Szabó, Thomas. 1997. "Die Kritik der Jagd, von der Antike zum Mittelalter." In *Jagd und höfischer Kultur im Mittelalter*, edited by Werner Rösener, 167–230. Göttingen: Vandenhoek & Ruprecht.

Tarr, Anita. 2000. "'Still So Much Work to Be Done': Taking Up the Challenge of Children's Poetry." *Children's Literature* 28:195–201.

Taylor, Joseph E. 1999. *Making Salmon: An Environmental History of the Northwest Fisheries Crisis*. Seattle: University of Washington Press.

Theimer, Sharon. 2009. "Promises, Promises: Is Obama Dog a Rescue or Not?" Breitbart.com, April 13. http://www.breitbart.com/article.php?id=D97HQQP00&show_article=1.

Thirsk, Joan. 1984. "Horses in Early Modern England: For Service, for Pleasure, for Power." In *The Rural Economy of England: Collected Essays*, edited by Joan Thirsk, 375–401. London: Hambledon Press.

Thomas, Elizabeth Marshall. 2000. *The Social Lives of Dogs: The Grace of Canine Company*. New York: Simon and Schuster.

Thomas, Keith. 1983. *Man and the Natural World: Changing Attitudes in England, 1500–1800*. London: Penguin Books.

Thompson, Wright. n.d. "A History of Mistrust: Having Trouble Seeing Why So Many Atlantans See the Michael Vick Case as Racial Conspiracy?" http://sports.espn.go.com/espn/eticket/story?page=vicksatlanta.

Torriero, E. A. 2002. "Kabul Zoo's Shell-Shocked Lion Dies." *Chicago Tribune*, January 27.

Trinkaus, Richard. 1983. *The Scope of Renaissance Humanism*. Ann Arbor: University of Michigan Press.

Tuan, Yi-Fu. 1984. *Dominance and Affection: The Making of Pets*. New Haven, Conn.: Yale University Press.

Turner, Terence. 1991. "'We Are Parrots,' 'Twins Are Birds': Play of Tropes as Operational Structure." In *Beyond Metaphor: The Theory of Tropes in Anthropology*, edited by James W. Fernandez, 121–158. Stanford, Calif.: Stanford University Press.

Tyler, Tom, and Manuela Rossini, eds. 2009. *Animal Encounters*. Boston: Brill.

Van Loo, Elizabeth. 1996. "A Study of Pet Loss and Its Relationship to Human Loss and the Valuation of Animals." Ph.D. diss., Troy State University.

Veblen, Thorsten. 1979. *Theory of the Leisure Class: An Economic Study in the Evolution of Institutions*. New York: Penguin.

Vint, Sherryl. 2007. "Speciesism and Species Being in *Do Androids Dream of Electric Sheep?*" *Mosaic* 40, no. 1: 111–126.

Vint, Sherryl. 2008. "'The Animals in That Country': Science Fiction and Animal Studies." *Science Fiction Studies* 105. http://www.depauw.edu/sfs/abstracts/a105.htm#vint.

Visser, A. S. Q. 2005. *Joannes Sambucus and the Learned Image: The Use of the Emblem in Late-Renaissance Humanism*. Leiden: Brill.

Viveiros de Castro, Eduardo. 1992. *From the Enemy's Point of View: Humanity and Divinity in Amazonian Society*. Chicago: University of Chicago Press.

Waddell, Helen, ed. and trans. 1934. *Beasts and Saints*. Woodcuts by Robert Gibbings. London: Constable.

Waller, Richard. 2004. "'Clean' and 'Dirty': Cattle Disease and Control Policy in Colonial Kenya, 1900–40." *Journal of African History* 45:45–80.

Waller, Richard and Kathy Homewood. 1997. "Elders and Experts: Contesting Veterinary Knowledge in a Pastoral Community." In *Western Medicine as Contested Knowledge*, edited by Andrew Cunningham and Birdie Andrews, 69–93. Manchester: Manchester University Press.

Wenner, Kathryn S. 2002. "Dog Gone, Not Forgotten." *American Journalism Review* 24, no. 4: 14–15.

White, Luise. 1995. "Tsetse Visions: Narratives of Blood and Bugs in Colonial Northern Rhodesia, 1931–9." *Journal of African History* 36:219–245.

White, Richard. 2003. "Tempered Dreams." In *Inventing for the Environment*, edited by Arthur Molella and Joyce Bedi, 3–10. Cambridge, Mass.: MIT Press.

Whitehead, Alfred North. 1930. *Science and the Modern World*. Cambridge: Cambridge University Press.

Willard, Charles Arthur. *Liberalism and the Problem of Knowledge: A New Rhetoric for Modern Democracy*. Chicago: University of Chicago Press, 1996.

Willerslev, Rane. 2007. *Soul Hunters: Hunting, Animism, and Personhood among the Siberian Yukaghirs*. Berkeley: University of California Press.

Williams, R., and D. Edge. 1996. "The Social Shaping of Technology." *Research Policy* 25:856–899.

Wilson, Robert A., ed. 1999. *Species: New Interdisciplinary Chapters*. Cambridge, Mass.: MIT Press.

Wolfe, Cary. 2003. *Animal Rites: American Culture, the Discourse of Species, and Posthumanist Theory*. Chicago: University of Chicago Press.

Wolfe, Cary. 2003. "In the Shadow of Wittgenstein's Lion: Language, Ethics, and the Question of the Animal." In *Zootologies: The Question of the Animal*, edited by Cary Wolfe, 1–58. Minneapolis: University of Minnesota Press.

Wolmer, William L. 2003. "Transboundary Conservation: The Politics of Ecological Integrity in the Great Limpopo Transfrontier Park." *Sustainable Livelihoods in Southern Africa Research Paper No. 3*. Brighton: Institute of Development Studies.

Wolmer, William L. 2007. *From Wilderness Vision to Farm Invasions: Conservation and Development in Zimbabwe's South-east Lowveld*. London: Routledge.

Wolter-von dem Knesebeck, Harald. 1997. "Aspekte der höfischen Jagd und ihrer Kritik in Bildzeugnissen des Hochmittelalters." In *Jagd und höfischer Kultur im Mittelalter*, edited by Werner Rösener, 493–572. Göttingen: Vandenhoek & Ruprecht.

Woods, Michael. 2000. "Fantastic Mr. Fox? Representing Animals in the Hunting Debate." In *Animal Spaces, Beastly Places: New Geographies of Human-Animal Relations*, edited by Chris Philo and Chris Wilbert, 182–202. New York: Routledge.

Wright, Allan. 1972. *Valley of the Ironwoods: A Personal Record of Ten Years Served as District Commissioner in Rhodesia's Largest Administrative Area, Nuanetsi, in the South-Eastern Lowveld*. Cape Town: T. V. Bulpin.

Ziolkowski, Jan M. 1993. *Talking Animals: Medieval Latin Beast Poetry, 750–1150*. Philadelphia: University of Pennsylvania Press.

Contributors

Sharon Wilcox Adams is a doctoral candidate in the Department of Geography and the Environment at the University of Texas at Austin. Her master's research addressed the construction, representation, and performance of race and ethnicity within the Afro-indigenous communities in Belize, culminating with her thesis, "Reconstructing Identity: Representational Strategies in the Garifuna Community of Dangriga, Belize." Her dissertation, "Encountering El Tigre: Jaguars and People in the United States, 1800–2010," examines the modern perception and representation of jaguars in the American Southwest.

Benjamin Arbel is professor of early modern history at Tel Aviv University. He has published seven books and about sixty articles dealing with Renaissance culture, intercultural contacts in the Mediterranean world, social and economic aspects of the Renaissance period, and historical aspects of human-animal interaction. The last include studies on Renaissance attitudes to animals, on dealing with locust in Venetian Cyprus, on the "triumph of the mule" in sixteenth-century Cyprus, and on the Venetian supply system of hunting birds. He is coeditor of *Human Beings and Other Animals in Historical Perspective*, published in Hebrew in 2007.

Etienne Benson is a research scholar at the Max Planck Institute for the History of Science in Berlin. He received his Ph.D. from the Massachusetts Institute of Technology (MIT) in 2008 and was a postdoctoral fellow at the Harvard University Center for the Environment from 2008 to 2010. He is the author of *Wired Wilderness: Technologies of Tracking and the Making of Modern Wildlife*.

Jane Desmond is professor of anthropology and of gender/women's studies at the University of Illinois at Urbana-Champaign, where she also directs the International Forum for U.S. studies, a center for transnational studies of the United States. She holds a Ph.D. from Yale University in American studies, and is the current president of the International American Studies Association (2009–2011). Her academic work focuses broadly on issues of embodiment and social identity, in the arenas of performance, visual culture, and tourism. Her books include *Staging Tourism: Bodies on Display from Waikiki to Sea World* and her current animal studies book project, "Displaying Death/Animating Life."

Avigdor Edminster holds a degree in English as well as one in historical studies, and completed his Ph.D. in anthropology at the University of Minnesota. His interests include multispecies relations and the mutual transformation possible in the interplay of ethnography and ethology, as well as anthropology of the senses. Avigdor is also a visual artist and sometimes dormant poet.

Linda Kalof is professor of sociology at Michigan State University and author of *Looking at Animals in Human History*. She is general editor for three multi-volume book series: *Cultural History of Animals* (winner of the 2008 *Choice* Award for Outstanding Academic Title), *Cultural History of the Human Body*, and *Cultural History of Women* and co-editor of *The Animals Reader*.

Clapperton Chakanetsa Mavhunga is a tenure-track assistant professor at Massachusetts Institute of Technology's (MIT) Program in Science, Technology, and Society. He is a theorist and historian of mobility in general and a historian of African mobilities and mobility in Africa. He has just completed his latest publications, "A Plundering Tiger with Its Deadly Cubs? The USSR and China as Weapons in the Engineering of a 'Zimbabwean Nation,' 1945–2009," in *The Technopolitical Shape of Cold War Geographies* and "Vermin Beings: On Pestiferous and Human Game," and is finalizing his first book.

Georgina M. Montgomery is an assistant professor with a joint appointment in Lyman Briggs College and the Department of History at Michigan State University. Her publications and book manuscript (in progress) explore the history of animal behavior studies, with particular focus on the people, places, and practices involved in the science of primatology.

Casey R. Riffel is a Ph.D. candidate in critical studies at the University of Southern California (USC) School of Cinematic Arts. His dissertation, titled "A Line of Escape: Animation, Animals, Modernity," explores the representation of animals in early animation in order to retheorize the medium's relationship to concepts of life and movement. His research and teaching interests include visual culture, early cinema and protocinematic devices, the history and theory of animation, animal studies, and the history and theory of technology. Before coming to USC as a Javits Fellow, he studied modern thought and literature at Stanford University.

Meisha Rosenberg is an independent scholar and writer specializing in arts and culture whose work has appeared in the International Journal of Comic Art, Journal of the Short Story in English, Bitch, Salon.com, the Albany (N.Y.) Times Union, and the Women's Review of Books. She is a graduate of Reed College and New York University and has taught at Siena College and the College of Saint Rose. The essay here came out of research for a book about perceptions of dogs in American culture and history.

Stacy Rule is a fourth-year Ph.D. candidate in the American studies program at Michigan State University (MSU). She recently completed a graduate specialization in animal studies: humanities and social science perspectives through MSU's Department of Sociology. She is a member of the Association for the Study of Literature and the Environment and has presented papers on human and animal relations at conferences in the United States and Canada. She lives in Hong Kong, where she is studying for comprehensive exams in American modernism, ecocriticism, and identity politics.

Analía Villagra is a doctoral candidate in cultural anthropology at the Graduate Center of the City University of New York. She is working on her dissertation on the intersection of land rights and conservation politics. Based on over a year of fieldwork in southeastern Brazil, her research explores diverse perspectives on nature and the environment among nongovernmental organizations, municipal governments, and social movements brought together by efforts to save an exceptionally rare primate, the golden lion tamarin. In addition to her graduate work, she has served as a conservation educator at the Prospect Park Zoo and Staten Island Zoo.

Index